T0245161

Food and fuel

Food and fuel

The example of Brazil

Marcos Fava Neves

Mairun Junqueira Alves Pinto

Marco Antonio Conejero

Vinicius Gustavo Trombin

Wageningen Academic
Publishers

Wageningen Academic Publishers
P.O. Box 220
6700 AE Wageningen
The Netherlands
www.WageningenAcademic.com
copyright@WageningenAcademic.com

ISBN 978-90-8686-166-8
e-ISBN: 978-90-8686-721-9
DOI: 10.3920/978-90-8686-721-9

First published, 2011

© Wageningen Academic Publishers
The Netherlands, 2011

Table of contents

Introduction

The objective of this book is to analyse the impact of transnational companies in Brazilian agricultural production, primarily in the sugarcane sector. To do this, the book first gives an overview of Brazilian agribusiness in Chapter 1, followed by the Brazilian Sugarcane Value Chain in Chapter 2. Chapter 3 focuses on the Investments of Transnational Companies (TNCs) in this sector in Brazil, and finally Chapter 4 concludes with lessons learned and a proposed agenda for public and private policies.

1. The need for food and fuel production and the role of Brazilian agribusiness

This chapter is divided into six main sections. The first section looks at the global financial and economic crisis and the impacts on the food sector and emerging nations. The second section focuses attention on the problem of increasing food demand and, as a consequence, rising food prices. The third section is a proposal on how society should work together to try to solve the growing food problem. Then, in the fourth section, we present a background and overview of Brazil's agricultural production. The fifth section highlights the importance of agribusiness for the Brazilian economy. We finish in the sixth section with a scenario for TNC investments, along with some trends and global issues.

1.1 The global economic/financial crisis

From a very simplistic perspective, the economic crisis of 2008/2009 was a crisis of the 3 Cs. The first C stands for *credit*. Credit was given to consumers in recent years in a very irresponsible way by financial institutions in many countries, and an artificial market was created. It was an era of financial leverage, financial strategy, and financial dominance. Numerous companies joined the party, building up very risky positions, paying out unrealistic salaries and dividends, and neglecting costs. When the festivities began to turn sour, adjustments were needed.

The second C stands for *consumption*. Society consumed in an irresponsible manner, also in many countries. The abundance of available credit meant that a great many consumers bought things they simply could not afford, taking out loans for houses and cars, etc. It was blatantly obvious that most families did not have the budget to cover this increased consumption. But the party continued. Now it is time to reduce leverage, to sell what was bought, at a massive loss, because the price of assets (houses, cars, etc.) has fallen.

The third C stands for *confidence*. The first two Cs caused society to lose confidence in the system, in companies, and even in governments. Economic recovery is related to confidence. It will not be easy and will vary from one country to the next. At the beginning of the crisis, there seemed to be no end in sight, so some consumers with purchasing power lost confidence and stopped consuming. With lower sales, markets shrink, employment falls, and this leads to lower consumption, lower sales, and unemployment with a negative cyclical effect, bringing deeper crisis. The speed of 'confidence recovery' is what is going to pull economies out of this crisis. The economic situation will get better way before the financial situation, because it is still unclear what will happen in terms of bad credit. In any case, the risk of a total financial collapse was mitigated.

Together with the economic and financial crises, the world in 2009 faced an escalating political crisis, due in part to governments that took the wrong political and economic measures, taking countries to the edge of bankruptcy, governments being removed, and an increasing threat of instability in the form of nuclear weapons and other forms of 'passive aggression', like missile testing and arming nations. This fact also disturbed markets and consumer confidence and formed part of the environment in which food chains operated.

It is very important to realise that the 2009 global crisis cannot be generalised. It had very different regional, country and industry impacts. As an example, some industries in Brazil were faced with record sales, in food, cars and housing, whereas some industries, like heavy equipment, suffered the worst crisis in 20 years. Some areas within countries and some cities suffered more, and others less. Some countries suffered more (Germany, with a 5% decrease in GDP) and some less (China, with a 6% increase). The USA continues to suffer due to high consumer leverage. Predictions for European recovery are worse than the Americas and Asia.

Another factor that contributed to a faster recovery in the economic situation after the crisis was the enormous shift in market power in the world. To a certain extent the emerging economies have already changed the world. According to Goldman Sachs, the BRIC (Brazil, Russia, India and China) group is expected to have a higher GDP than the G7 in 2027. Between 2000 and 2005 the BRIC GDP rose from US$ 3.6 trillion to almost US$ 5 trillion. Between 2000 and 2009 Brazil experienced an average growth of 3.1% per year, and an accumulated variation of 36.3% in the same period. China experienced an incredible 9.6% average and 151% accumulated growth. India is witnessing an average 6.9% and 94.6% accumulated growth, and finally, Russia saw 5.5% average growth (2000/2009) and 71.2% accumulated (IMF, undated).

The GDP of the emerging nations was 11% of the world in 1991, 30% in 2008 and is expected to be 50% in 20 years. The global population is expected to reach 9 billion in 2050, and only 10% of that will be in the developed nations. In 2009, there were around 200 million people in emerging nations with an income of US$ 3,000 per year; this will rise to 2 billion people in 15-20 years. So there is no question that a huge shift has occurred in the last 10 years. New consumers and new markets are diversifying the world. This shift will gain more and more speed; it will be accelerated by the effects of the internet with faster technology and knowledge transfer.

The GDP of large food consumers in nations like China and India continues to grow, contributing to maintained or even increased consumption. Goldman Sachs expects China to grow 7-8% in 2009 and 11% in 2010, and India 6% to 7% in 2009/2010. Almost 60% of the world's economic growth in the period 1999/2009 was in the developing nations, namely 30% in the BRIC. The World Bank estimated that in 2009 the global economy would shrink by 2.9%, and then grow by 2% in 2010, and 3.2% in 2011. Half of this growth would come from the emerging nations and markets.

Those countries that are heavily dependent on the USA, for imports or tourism, and on flows of money coming from populations working abroad, are suffering more than the traditional food and commodity producers and exporters, like Brazil, Uruguay, Colombia, Argentina and others. Brazil is well known as a major food and biofuels exporter. Food is more resilient to crisis since this commodity is the last thing to be cut from a family budget, and, in the developed world, food demand elasticity is low. The markets in the higher value sectors were hit harder.

For various reasons, this crisis is not having the same effect on Brazil as in the past. This is partly because of the Real Economic Plan, launched in 1994, that brought economic stability. As an example, supermarket sales in April 2009 were 6.5% higher than April 2008. The economy has been growing around 4-5% per year in the last few years, and will probably fall to +0.5-1% in 2009. There is a general agreement that it will be back to 3-4% growth in 2010. Goldman Sachs expects a 5% increase. The country has a small share of general global trade (it is only important in food and commodity trade) and was less affected than other export-driven economies. A large domestic consumption market, tough adjustments made in the past in the banking and financial system on credit exposure, a high level of international reserves (stable at US$ 205 billion – other countries like Russia, Mexico and Korea lost reserves), a situation of energy security (self-sufficiency in oil and with more than 50% of car fuel nowadays coming from domestically produced sugarcane ethanol) and other factors contributed to this position. As recently as May 2009 Brazil received US$ 2.3 billion in direct foreign investment. The accumulated figures for 2009 are US$ 11.2 billion.

We conclude this first section with a question: what are the messages for companies? There are several. Companies will need to start focusing again, to return to their core business, make very efficient use of capital and resources, and work even more on planning, collective actions and cost structure. Companies will also need to take a very close look at risk monitoring. It is an era of establishing global and more competitive supply chains and an era of strong value proposition for human talent in companies. Finally, it is an era of more conservative leverage and financing, and the chance to take advantage of opportunities for consolidation, acquisition, and mergers, etc., as well as for cheap asset acquisition in the world. Those businesses with capital in 2009 had an outstanding window of opportunity.

1.2 Food demand and food inflation

The second section of this chapter deals with a problem that arose in 2007 and part of 2008 and will recur sooner than expected for various reasons. This problem is increased food demand and as a consequence, the rise in food prices.

In 2007 and 2008 there were several important discussions in international forums regarding booming food prices and its consequences worldwide. Between 2003 and 2005, the FAO's food price index rose 14.71%; in the subsequent two years, it reached 34.19%; then in just one year from March 2007 to March 2008, the index jumped to an incredible 57.14% (FAO, 2008a). As a result, inflation was a real concern in Europe (3.6%), China (8.3%), USA (4.0%), Russia (12.7%) and many other markets of varying economic power. The poorest countries suffered the most from soaring food prices, because poorer families spend a larger percentage of their budget on food consumption.

There was devastating proof of this in several of the least developed nations during 2008. For example, in Haiti, the poorest country in the Americas, former Prime Minister Jacques Edouard Aléxis was expelled from his position by his own congress after being accused of negligence in failing to properly combat the problem of food prices. Within one week, in Africa, a fifty kilo bag of rice went from $ 35 USD to $ 70 USD. In Cameroon, Africa, the official numbers of dead were reported to be 24 after weeks of conflicts between local police and hungry mobs, but some human right activists say the real number exceeds the hundreds. According to FAO Director, General Jacques Diouf, soaring food prices pushed no less than 50 million people into hunger in 2007 alone (FAO, 2008b).

In a speech made in 2008, UN General Secretary Ban Ki-moon asked for food production to be doubled by 2030. According to FAO, even with the reduction in hunger in the world (between 1969-2000 the numbers of starving people fell from 37% to 15%), food prices in some countries were 80% higher in 2009 than in 2007. On average, food prices are 24% higher than in 2006. FAO also expects commodity prices to rise by 10-30% in the next 10 years, claiming that food is not a priority in global politics and that this should change.

Following the first news about rising food prices and its consequences, journalists, researchers, scholars and opinion-makers started to publish their studies and personal thoughts about the causes of these dynamics. There are 9 major factors that are changing and putting pressure on the ability to supply food to the world, and are related to the economic and financial crisis.

1. Increase in areas dedicated to growing crops for *biofuels*: several countries are starting to produce biofuels, and in some of them this is taking up land previously used for food production. The car is now competing with humans for food. Many studies are only linking biofuels to the food inflation factor, ignoring several other factors, some of which we've known about for a long time. Biofuels are not the main problem, since there are very positive results of biofuels being produced in areas together with an increase in food production, but this factor should be considered.

2. The *growth of the world population*, expected to reach 9 billion people in 2050, is not a new factor, but it contributes to a need for increased food production. FAO/ONU estimates that the world will need to produce at least 50% more food in the next 15 years. Projections of future demand for grains (2.2 billion in 2009 to 3.3 billion in 2025), milk (3.4 billion tonnes in 2009 to 5 billion tonnes in 2017) and meat are impressive. Just as an example, the MENA (Middle East and North African) countries have a population of around 380 million today, and this will rise to 510 million in 2025.

3. Economic development and *income distribution in populated countries* such as India, Brazil, Eastern Europe, China, Indonesia, Thailand, South Africa, Argentina, Arabic and African countries, among others, are among the most exciting factors, bringing millions of new food consumers to the market. Several African economies have grown more than 5% per year in the last 5-10 years. In the next ten years, experts in food consumption expect an increase in food expenditure to the order of 50% in China, 78% in India, and 40% in developing Asia, Middle East and North Africa (Global Demographics Report, 2008). The 60:40 consumption ratio of developed vs. emerging economies will become 50:50 within a decade. When one compares China's population with the country's participation in world trade, it is still less than 50%. There is a lot more to come.

4. Stronger *governmental programmes* for aid and food consumption, such as the one introduced in Brazil reaching 10 million families and 40 million people, bringing new consumers coming to the food markets. Just as an example, the market for sausages in Brazil increased from US$ 0.5 billion in 2003 to US$ 1 billion in 2007. Furthermore, Thailand has 10 million people receiving a cheque for US$ 58 per month. These are just some examples occurring in several parts of the world, and just some of the many signals that are not being adequately picked up by major economists.

5. *Migration and urbanisation* of communities creating mega-cities, increasing food consumption and changing consumption habits favouring less grain and more protein (using more grain as feed in the animal production process); consumption is becoming more individually based, more sophisticated and more energy consuming. There has also been a huge impact here, when you consider that in some countries 50% of the population still lives in rural areas and is moving to the cities. A study carried out by McKinsey (Aston, 2009) estimates that around 350 million people in China will move to the cities by 2025. This will require 5 million buildings, equipped with computers, televisions, air conditioning and will result in new food consumption habits (the equivalent of 10 cities the size of New York).

6. *Oil prices* went up from $ 35 USD to $ 140 USD in five years, affecting production and transport costs. It is rising again, and oil is not just used for transportation. It is also used in several other industries, like plastics, that have increased their consumption. Oil may be stable around US$ 70-80 a barrel, and with oil prices up again, several possibilities and projects for biofuels are gaining economic incentives, increasing the pressure for land in the case of maize and other grains. China had 65 million cars on the road in 2008 and is expected to have 150 million by 2020, consuming 250 million tonnes of gas/year (Xinlian, 2009).

7. The *dollar devaluation* that happened in recent years also played its part in higher commodity prices, fixed in US$.

8. *Production shortages (food supply)*. Farm/production shortages due to lower margins, climate, droughts and diseases are a major concern. |The credit crunch (lack of financing) and massive price fluctuations caused a downturn in prices and financial institutions tightened their criteria. All this, together with losses from bad hedging by agribusiness companies, brought about a loss of confidence. As a consequence, there were higher risks

for planted area and yields, hedging prices got worse (less hedging), leading to more uncertainty and a lack of confidence in long-term contracts. This may bring with it lower productivity, lower inventories, lower margins and cause farmers to switch to producing cheaper crops. Some exporting countries will become importers and the 2009 crop was predicted to be smaller in several countries (expected 5% smaller global production). On this same topic, a major point of concern is the availability of water and the cost of water, as well as the unknown impacts of global climate change on crop productivity in the future.

9. *Investment funds* operating in futures markets and others in agribusiness. This increased due to the lower interest rates in many countries. It is know that some of these have been replaced by strategic investors with conservative financing mechanisms, but there is still a movement of funds towards food commodities, which is also increasing consolidation.

To what extent each of these factors is responsible for causing the problem is an area for future research. However, the most important thing to do is monitor these factors and see whether they trigger the problem again.

In the particular case of biofuels, several important sustainably-driven global investments in South America, Africa and Asia, among others, are being severely damaged by these articles and opinions. Biofuel production in Brazil is more than 1,500 miles away from the Amazon region. An important representative from the United Nations (ONU) has classified biofuels as a 'crime against humanity' and has requested that the European Commission abandon its target of blending fossil fuels with 10% biofuels. The General Director of Food Marketing Institute (FMI) has attacked biofuels, classifying their production as a 'moral problem'.

Several significant research studies have been published validating positive experiences and solutions regarding the sustainability of biofuels for decades and must be considered before voicing an opinion. Since the debate is gaining ground, we must scrutinise studies that are being published in front-line world journals, newspapers and magazines, using sometimes obscure methodologies and dangerously generalising the results. Academics know the risks of generalisation.

Unfortunately, not all biofuels can be put in the same basket, because there are significant differences among ethanol sources and their energy-yield efficiency. Brazil manages to produce its sugarcane-based ethanol without subsidies, for less than half the cost per litre and more than twice the yield per hectare compared to the US product. South-central Brazil produces around 7,500 litres per hectare, while the ethanol produced in Europe with sugar beet ethanol yields around 5,500 l/ha and USA maize ethanol yields around 3,800 l/ha.

As far as land use is concerned, sugarcane has mostly taken over areas of degraded pasture used for extensive cattle farming. In the state of Sao Paulo, where currently around 70% of the country's sugarcane grows, the area designated to the crop grew more than 37% from

2001 to 2006. Some 75% of such growth, or 725,204 hectares, occurred on former pasture areas. During that period, pastures also lost ground to soy, sorghum, cassava, potatoes and other important crops for human and animal nutrition (Camargo, 2007).

Brazil's livestock index is 1 animal per hectare, significantly lower than other international meat exporters, which leaves a huge gap for improvement in yield. Nevertheless, modernisation is already taking place on Brazilian ranches through investments especially in genetics and animal nutrition, as well as improvements in animal and soil handling.

In addition, because of the need for crop rotation, 15 to 20% of the areas used for growing sugarcane are actually producing food (usually soy, peanuts or beans). This has contributed to Brazil's record food production year after year, despite the increase in biofuel production. As mentioned before, there are several studies showing that Brazilian ethanol and other biofuels are energy and cost-efficient and represent a sustainable path towards the development of some of the world's poorest areas. Society should ask itself what the interests are behind these 'studies' and who is sponsoring them. A good starting point is to analyse who loses margins with these changes and the growth of biofuels.

1.3 The road to addressing increased food demand and food inflation

In the third section of this chapter, we outline two ways of solving this potential food demand/inflation problem. One is to go back to an increase in protectionism, stimulating non-competitive areas to produce in an 'economically artificial environment' and returning to policies of war-time 'self-sufficiency'. The other is to move forward towards growth, global trade and inclusion. This is presented as a 10-point agenda which can be used by governments and international organisations as an outline for possible solutions to the upcoming food demand and inflation problem. This agenda could ensure long-term results, peace, income distribution and inclusion. Each point will be addressed in sequence:

1. *Promoting horizontal expansion in production into new areas, with environmental sustainability.* This expansion can be undertaken in several countries (South America uses only 25% of its capacity), on all continents, in millions of hectares that today are under-utilised. In Brazil several studies by recognised institutions confirm the existence of more than 100 million hectares that can be used for food and biofuel production, without affecting fragile systems and mostly growing over degraded pastures. This production and land expansion, if stimulated by sustainable contracts, encompasses farming, new entrepreneurs, job creation in less developed nations, income distribution and economic development, and even has a positive impact on democracy. Land costs are rising, because numerous pension funds are looking for security and buying up land. Recently a fund of US$ 800 million in Arabian countries was dedicated to land purchase and food security, with South American and African countries as targets (Financial Times, 2009). There have been several moves by China and other countries to build

these supply chains abroad. This is a perfect match between investments and the need for development.

2. *Promoting vertical expansion, or more production in areas that are already being utilised.* Many hectares in South America, Africa, Asia, and even in developed nations could produce more if there was more investment in technology. As a comparison, the amount of maize a USA farmer can generate in tonnes per acre is two or even three times higher than the average production in Brazil and other countries. With irrigation systems, some farms in the tropics could generate three crops per year. Major research and investment should be dedicated to these improvements.

3. *Reducing food import taxes and other import barriers and protectionist measures.* Food prices in some countries are artificially inflated due to import taxes and other kinds of protectionist measures that damage international trade, markets and growth. As an example, beef in the European Union costs four to five times more than the same quality of beef in an Argentinean or Brazilian branch of the same European retailer. The argument most widely used is that lowering protection barriers will damage local agriculture in less developed countries. It must now be assumed that the new level of commodity prices may allow local agriculture to be competitive. Several other internal taxes on food could also be reduced by local governments, reducing consumer prices. In addition, the more than $ 330 billion USD spent annually by OECD members on agricultural subsidies puts more pressure on prices while undermining more cost-efficient food production in naturally competitive countries. The 2009 crisis may mean that money and funds are withdrawn from subsidies in developed nations, since these are not high on the list of groups to receive national government funds in times of crisis.

4. *Investing in international logistics in order to reduce food costs.* Some grain-producing countries have extremely poor logistics. Governments should invest and society should work harder to change institutions in order to facilitate public private partnerships to privatise ports, roads, and other food distribution and logistic systems to make the flows faster and more energy-efficient. The losses here are great.

5. *Reducing transaction costs.* Major international food chains are affected by poor coordination and communication, ineffective administration, poor use of assets, corruption, opportunism and other inefficiencies that are largely responsible for losses, increased costs, and the maintenance of companies that don't add value, and agents or others that have an impact on food prices. Institutional reforms as proposed by Douglass North are the solution here. Also, more efficient cooperatives, producer pools, and other collective actions should gather force to reduce inefficiencies and increase producer organisation and bargaining power.

6. *Using the best sources for biofuels,* in a totally sustainable way. The example of Brazil could be better analysed, since ethanol has been produced for more than 35 years here, with 3.5 million hectares of cane, using only 1% of the country's arable land and supplying 52% of fuel transport consumption, with no impact on food production. The simultaneous growth of food production and biofuels in the State of Sao Paulo (the major area of sugarcane growth) in the last 10 years, shows that it is possible to grow crops for biofuels

that have better yields and don't compete with food chains. This should be prioritised in the global development of biofuels. The energy balance of sugarcane ethanol is 4.5 times better than that of ethanol produced from sugar beet or wheat, and almost seven times better than ethanol produced from maize (Figure 1).

7. *Investing in a new generation of fertilisers.* It is important to produce fertilisers from alternative sources, plants that can absorb more solar energy, with more recycling of by-products as sources of fertilisers to mitigate the huge risk and cost of fertilisers in the future. Fertilisers are among the most important and expensive inputs for agriculture, and at times when yield must be improved, its importance is even greater. In the last three years farmers have been confronted with a dramatic increase in the price of fertilisers. Compared to 2006, the DAP international price averages for the first four months of 2008 rose more then 360%. In addition, during this same period, the price of phosphate rock rose by 730%, TSP by 414% and urea and potassium chloride by 181% and 197% respectively (Figure 2).

8. *Working more towards sustainable supply contracts for farmers, with integrated sustainable investments and projects.* It is of fundamental importance that margins and income are better distributed along food chains, reaching farmers all over the world. Price stimulus is the best economic incentive for growth in production with technology. The idea that concentration in a handful of food industries and retailing businesses maintains margins that could be better distributed to farmers, increasing economic development and bringing a very positive externality to several regions, has been well researched. Models of integrated projects with sustainable development could be used (Neves and Thomé e Castro, 2009).

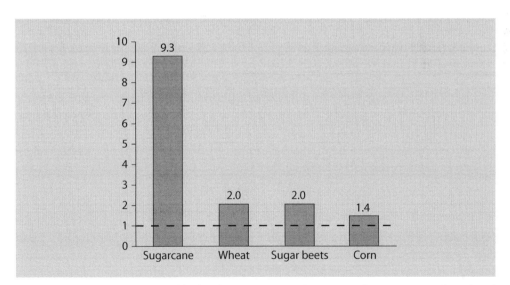

Figure 1. Energy balance by type of feedstock. Data represent the amount of energy contained in ethanol per unit of fossil fuel input (World Watch Institute, 2006 and Macedo *et al.*, 2008). Estimated data compiled by Icone and Unica.

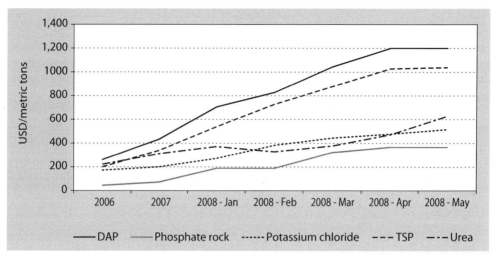

Figure 2. International fertiliser prices (World Bank, 2008).

9. *Stimulating research and investment in innovation from all possible sources, but mostly in genetics, in order to find new solutions for food and biofuel production and consumption.* In trying to solve the sustainability equation, the problem today is the shortage of seeds. Public investment in agricultural research and development has decreased considerably in the past couple of decades, resulting in a yield-growth slowdown, disabling production and the ability to keep up with rising consumption. Since trust in biotechnology is increasing in society, bringing a new era of acceptance, research should receive more attention.

10. *Gradually working to change consumption habits in both food and fuel.* We must realise that there aren't sufficient resources on the planet to allow 9 billion people to live the lifestyle of developed nations. Behaviour should gradually be changed towards sustainability. Food is over-consumed in many parts of the world, bringing with it obesity – a major health concern. Another area of inefficient consumption is fuel. Investments need to be made in resourceful public transport. This is a major challenge in many countries. Barcelona has implemented a very nice public biking system which is an excellent example of a working solution. Table 1 resumes the discussion, bringing together all the causes and the proposed solutions.

These 10 suggestions are not new, and some are already being implemented with good results. But there comes a turning point. We can either go back to trying to increase protectionism, undertake less efficient ventures in self-sufficiency, ban biofuels, create food export taxes, and even threaten to turn private companies into public companies. Or society can move forward, with this positive agenda, which is the right road for global sustainability.

Table 1. The food demand model.

9 causes of food price increases	10 proposed solutions
Biofuels	Sustainable horizontal expansion towards new areas
Population growth	Vertical expansion with more technology (high tech)
Income distribution and wealth in populated countries	Reduction in food taxes and other protectionist measures
Governmental programmes for food distribution	Investment in international logistics platform
Urbanisation of population and mega-cities	Use the best sources for biofuels production
Oil price impact on production and transportation costs	Reduction in transaction costs in food chains
Production shortages due to adverse climate and financial conditions, water and climate change impacts	New-generation fertilisers
Dollar devaluation	Sustainable supply contracts to farmers
Investment funds operating in commodities	Innovations (genetics and others)
	Adapting consumer behaviour for lower energy consumption

In order to accomplish the first two solutions for a more stable food and biofuel system, direct foreign investments by transnational companies are of fundamental importance. In most areas where there is an increase in crop production and crop productivity, there is basically one major restriction: lack of capital, and this is one of the most important solutions that a TNC company operating at the farm level abroad may bring. The next section gives an overview of Brazilian agriculture and agribusiness, one of the major targets of TNC foreign direct investments.

1.4 Background and overview of Brazil's agricultural production

Covering 851 million hectares, Brazil is a country of continental proportions. Of all this land, 40% is considered to be arable, which means that 340 million hectares could be used for agriculture and livestock. The other 511 million hectares include the 80% of the Amazon Rain Forest area which is protected by law; other conservation areas such as river banks, wetlands and reforestation areas; cities and towns; roads; lakes and rivers. Table 2 summarises the land use in the country.

The agricultural crops which occupy the most land are pasture, soybeans, maize, sugarcane, common beans, rice and coffee, in that order. Other important crops in terms of area are cassava, wheat, cotton and orange. Some of these agricultural products are concentrated in specific regions, while other crops are diffused. Pasture, which occupies over half of the country's arable land, is mostly non-native and is mainly used for raising beef and dairy cattle. Soybean, maize and common beans are mostly concentrated in the central southern

Table 2. Land use in Brazil and expansion possibilities (IBGE, CONAB and MAPA; elaborated by GV Agro).

	Million hectares	% of total land	% arable land
Brazil	850		
Total preserved areas and other uses	510 (60%)		
Total arable land	340 (40%)		
Cultivated land: all crops	72.0	8.5%	21.1%
Soybeans	21.3	2.5%	6.3%
Maize	14.6	1.7%	4.3%
Sugarcane	7.8	0.9%	2.3%
Sugarcane for ethanol	4.7	0.5%	1.4%
Oranges	0.9	0.1%	0.3%
Pastures	172	20.2%	50.6%
Available land (agriculture, livestock)	96	11.3%	28.2%

region, although they are also found in the north-northeast (in particular, maize and common beans).

One of the reasons why these three cultures are so widespread is, of course, their profitability, but also the development of varieties adapted to different soils and climates, and the fact that these are annual crops which are commonly used for crop rotation. Cotton, on the other hand, is mostly planted in the central and north-eastern regions. Rice production is higher in the south and south-east of Brazil. Sugarcane production is highly concentrated in the central southern region of the country. There are other foci of production located on the north-east coast, north of Rio de Janeiro and Espirito Santo. In the north of Brazil the production of sugarcane and soybeans is minimal or almost non-existent. Figure 3 shows the distribution of some of this agricultural production within Brazilian territory.

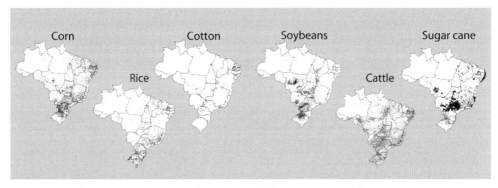

Figure 3. Distribution of agricultural products in Brazil (Source: PAM/IBGE).

Food and fuel

As can be seen, agricultural production takes places in virtually every region in Brazil. The first agricultural frontier expansion took place in the 1970s and 1980s. Based on tropical R&D, official rural credit and intervention prices, agriculture (mainly soybeans) expanded from the south of the country towards warmer savannah areas located in the centre. From 1970 to 1985, the area used for crop production increased about 53%, from 34 to 52 million hectares. In 1970, there were just over 1.3 million hectares planted with soybeans, which represented less than 4% of the total crop area. In 1985, soybean already covered over 10 million hectares and represented around 20% of the total area used for crop production. The total increase in the area planted with soybean reached 670% within those 16 years.

Other crops that experienced significant growth during the same period were oranges, which grew 228% reaching almost 0.7 million ha; sugarcane, which grew 127% reaching 3.9 million ha; common beans, which grew 53% reaching 5.3 million ha; and wheat, which grew 41%, reaching 2.6 million ha. Maize production grew somewhat less (20%), but was the most planted crop by 1985, covering 11.8 million ha. The only crops that shrank in area during that period were cotton (-16%), which suffered big losses due to pests, and rice (-5%) (ICONE, 2007).

The second expansion was in the 1990s and 2000s, pushed by greater demands and gains in efficiency (higher yield). There was little agricultural expansion compared to the previous period, with crops moving into the north and north-eastern regions of Brazil. From 1990 to 2005, of the most planted cultures, only soybean, sugarcane and maize experienced increases in planted areas. The area for soybean increased 100%, reaching 22.9 million ha by 2005.

The area planted with sugarcane expanded 36% within the same period, reaching 5.8 million ha. The area planted with maize stayed virtually the same, increasing only 1% to reach 11.5 million ha (ICONE, 2007). Currently, the great majority of Brazil's 96 million ha of available arable area is located in the 'cerrado' (savannah) and this region is slowly diversifying from the beef/soy model to maize, cotton, poultry, pork, sugarcane, dairy and coffee. Figure 4 shows the evolution of the ten most planted crops (90% of total crop area) from 1931 to 2003, in million hectares.

According to projections made by the Ministry of Agriculture, production of the main regional crops is expected to grow in all large producing states, as shown in Table 3. Even in those few cases where the planted area is expected to fall, total production should increase through gains in yield.

In addition to horizontal expansion (production increase through expansion in new areas), Brazilian agricultural production has benefited from major gains in vertical expansion (yield) in the last couple of decades. Data compiled by GV Agro, a Brazilian research centre, shows that between 1990/1991 and 2009/2010 Brazilian grain production increased by 144.1%. Meanwhile, the planted area increased by only 26.4%. One could therefore conclude

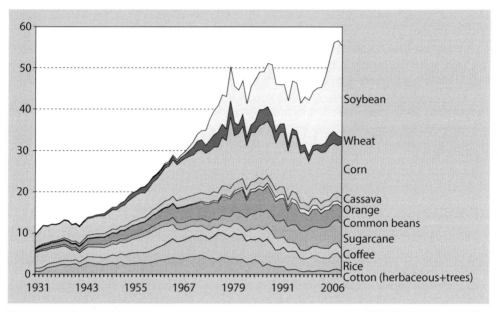

Figure 4. Evolution of the crop area in Brazil (1931 to 2006). Elaborated by ICONE based on IBGE – Estatísticas do Século XX, IBGE-Sidra e IPEADATA – séries Históricas.

Table 3. Projections of production and planted area for 2008/2009 and 2019/2020 (MAPA, 2010a).

	State	Production (thousand tonnes)			Planted area (thousand ha)	
		2008/2009	2019/2020	Var.%	2008/2009	2019/2020
Rice	Rio Grande do Sul	7,950	9,760	23.5	1,105	1,252
Maize	Mato Grosso	8,082	15,705	94.3	1,641	3,091
	Parana	11,101	16,675	50.2	2,783	3,258
	Minas Gerais	6,451	8,572	32.9	1,284	1,187
Soybean	Rio Grande do Sul	7,912	8,533	7.8	3,823	4,041
	Mato Grosso	17,963	27,944	55.6	5,828	8,289
	Parana	9,510	13,225	39.1	4,069	5,108
Wheat	Parana	3,201	3,769	17.7	1,152	1,138
	Rio Grande do Sul	2,059	2,553	24	980	815
Sugarcane	Sao Paulo	400,539	601,892	50.3	4,691	6,817
	Parana	55,086	90,280	63.9	644	860
	Mato Grosso	16,853	23,906	41.9	246	341
	Minas Gerais	56,098	98,155	75	679	1,129

that the use of high technology has resulted in a 93.2% increase in the average yield within that same period, proving the importance of research and development in techniques and technology (Figure 5).

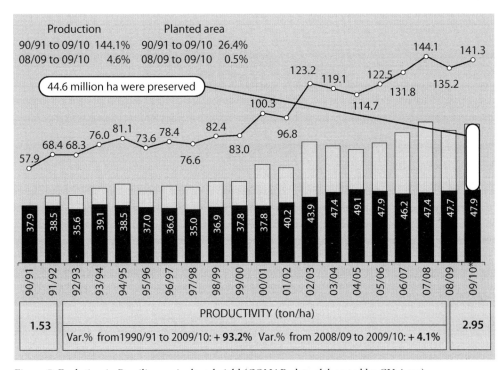

Figure 5. Evolution in Brazilian agricultural yield (CONAB, data elaborated by GV Agro).

Livestock production has also increased significantly. Between 1990 and 2009, chicken production increased by 376% (Figure 6), beef production by 83% (Figure 7) and swine production by 195% (Figure 8).

For the 2009/2010 harvest, the Strategic Planning Department of the Brazilian Ministry of Agriculture (AGE/MAPA) projects an increase in the production of grains after high input costs, especially fertilisers, led to a decrease in 2008/09 when compared to the previous crop year. As far as future projections are concerned, the Ministry expects a total production of almost 180 million tonnes by 2019/2020, as shown in Table 4.

Much of the boom in Brazilian agriculture production in the last couple of decades is due to an increase in the use of fertilisers. In 1983, the country had to import only 32% of total nutrients for agricultural production. In 2006, imported nutrients had already reached

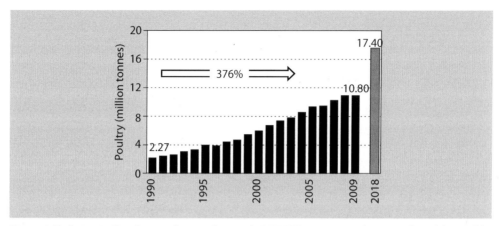

Figure 6. Evolution in Brazilian poultry production (ABEF, USDA, IBGE and MAPA, data elaborated by GV Agro).

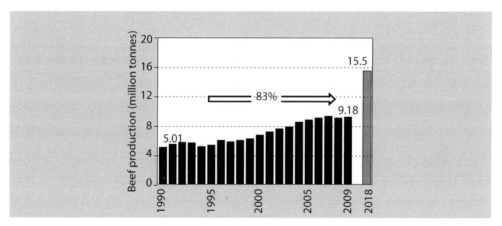

Figure 7. Evolution in Brazilian beef production (CNPC, ABIEC, USDA, IBGE and MAPA, data elaborated by GV Agro).

65%. Some estimates show that by 2025, Brazil will have to import 86% of total nutrients for agriculture (MB Agro, 2007). The greater demand for fertilisers allowed the increase in production, but has also increased production costs. Furthermore, producers are more vulnerable to fluctuations in fertiliser prices. In the 2008/2009 harvest, the production costs of soybean in the state of Mato Grosso rose 16.4% compared to the previous harvest. The price of fertilisers, which accounted for about 45% of total costs in the harvest, rose by 32.5% within one year.

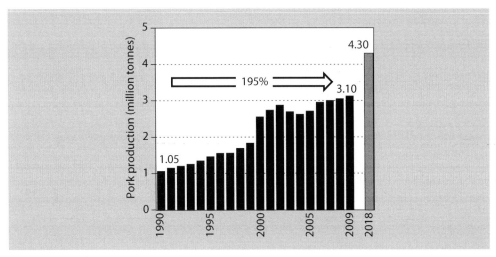

Figure 8. Evolution in Brazilian pork production (ABIPECS, USDA, IBGE and MAPA, data elaborated by GV Agro).

Table 4. Brazilian grain production (million tonnes) (based on MAPA, 2010a).

Crop	2008/2009	2009/2010	2019/2020
Cotton	1.19	1.27	2.01
Rice	12.63	12.59	14.12
Common beans	3.48	3.58	4.27
Maize	50.97	53.85	70.12
Soybeans	57.09	61.77	81.95
Wheat	5.67	5.09	7.07
Total	131.03	138.13	179.52

1.5 The importance of agribusiness for the Brazilian economy

Brazil is one of the most important exporters of agricultural products. It is the world's leading exporter of sugar and ethanol; chicken; beef; coffee; tobacco and orange juice. In the case of orange juice, Brazil holds 85% of the world's market share. Between 1999 and 2009, the volume of cereal exports increased by 35.9% and the volume of exported meat within the same period rose by 19.8%. Table 5 shows the market share and ranking of Brazil in some of the main agricultural products, as well as the annual growth rate between 1999 and 2009 and the volumes exported in the last crop year (2009/2010).

Table 5. Exports of Brazilian agricultural and livestock products (based on USDA, 2010 and MAPA, 2010b).

Product	Exported volume in 2009/2010 (million tonnes)	Brazil/World (2009)		Annual volume growth rates (1999-2009)
		Share	Ranking	
Soybean (grain)	28.53	39%	2	16.4%
Chicken	4	38%	1	19.8%
Beef	2.11	22%	1	19.8%
Pork	0.63	12%	4	19.8%
Orange juice	2.08	85%	1	3.1%
Maize	7.64	9%	2	-
Coffee	-	32%	1	5.7%
Cotton	0.47	12%	4	-

Such a significant performance has a major impact on the Brazilian trade balance. The last time the country experienced a negative trade balance was in 2000; since then agribusiness has been largely responsible for the consecutive positive balances. In 2009 Brazil exported US$ 152.2 billion, of which agribusiness exports amounted to US$ 64.7 billion (42%). Meanwhile, total imports reached US$ 127.6 billion, resulting in a positive balance of US$ 24.6 billion. However, if it hadn't been for the agribusiness balance of US$ 54.9 billion, the country would have finished the year with a deficit. The role of agribusiness in the trade balance in the last 15 years is represented in Table 6.

Agribusiness companies are among the country's biggest exporters. In 2009, Bünge Alimentos S/A exported US$ 4,343.9 million, falling behind only the oil giant Petrobras and Vale do Rio Doce, one of the world's largest enterprises in the mining industry. ADM's Brazilian subsidiary, with exports worth US$ 2,592.9 million, and the Brazilian unit of Cargill, which exported US$ 2,335.6 million, occupied the fifth and sixth position in the ranking of Brazilian export firms in 2009. Table 7 shows the largest Brazilian agribusiness export firms in 2008 and 2009, ranked among the firms of all sectors of the economy. Among the top 30 export companies are 14 agribusiness firms. Over half of them had negative growth compared to 2008 results, a situation partially explained by the world's recent financial crises, which have had a huge impact on world trade in general, and on exports to the developed world in particular.

One of the trends that has already started to affect Brazil's export of agricultural products is the ongoing sharing of economic power among countries. In 2005 the International Monetary Fund (IMF) estimated that G7 GDP would grow at an annual rate of 2.2% from 2006 to 2013, while the GDP of the developing countries would grow 7.1% per year during the same period. This means that the share of the G7 countries in the world's GDP would

Table 6. Brazilian trade balance from 1994 to 2009 (US$ billion) (based on SECEX).

Year	Brazil			Agribusiness			Others		
	Exports	Imports	Total balance	Exports	Imports	Agri balance	Exports	Imports	Others balance
1994	43.5	33.1	10.4	19.1	5.7	13.4	24.4	27.4	-3.0
1995	46.5	49.8	-3.3	20.8	8.6	12.2	25.7	41.2	-15.5
1996	47.7	53.3	-5.6	21.1	8.9	12.2	26.6	44.4	-17.8
1997	53.0	59.7	-6.7	23.4	8.2	15.2	29.6	51.5	-21.9
1998	51.1	57.6	-6.6	21.5	8.0	13.5	29.6	49.6	-20.0
1999	48.1	49.3	-1.2	20.5	5.7	14.8	27.6	43.6	-16.0
2000	55.1	55.8	-0.7	20.6	5.7	14.9	34.5	50.1	-15.6
2001	58.2	55.5	2.7	23.9	4.8	19.1	34.3	50.7	-16.4
2002	60.3	47.2	13.1	24.8	4.5	20.3	35.5	42.7	-7.2
2003	73.0	48.2	24.8	30.6	4.7	25.9	42.4	43.5	-1.1
2004	96.4	62.8	33.7	39.0	4.9	34.1	57.4	57.9	-0.5
2005	118.3	73.5	44.8	43.6	5.2	38.4	74.7	68.3	6.4
2006	137.5	91.4	46.1	49.4	6.7	42.7	88.1	84.7	3.4
2007	160.6	120.6	40.0	58.4	8.7	49.7	102.2	111.9	-9.7
2008	197.9	173.2	24.7	71.8	11.8	60.0	126.1	161.4	-35.3
2009	152.2	127.6	24.6	64.7	9.8	54.9	87.5	117.8	-30.3

decrease from 59.5% to 53.6%, while the share of developing countries would increase from 24.7% to 30.8% by 2013.

Along with this trend, developing economies have become increasingly important for Brazil's exports of agricultural products. According to data from the Ministry of Development, Industry and Trade, in 2000 around 30% of Brazilian agribusiness exports were destined to developing countries. From June 2007 to May 2008, this share reached almost 50%. Finally, with the impacts of the last financial crisis on the purchasing power of the developed economies, in the second half of 2008 the share of developing nations for Brazilian agribusiness exports overtook that of developed countries. In 2009 Brazil exported US$ 34.9 billion in agricultural products to developing economies and US$ 29.8 to developed countries.

Combined, the EU countries form the biggest market for Brazilian commodities exports. Nevertheless, from 1999 to 2009 its share decreased from more than 40% to less than 30%. In 1999, the USA was the second biggest buyer of Brazilian agricultural commodities, with

Chapter 1

Table 7. Top agribusiness export companies 2008 to 2009 (US$ million) (Balança Comercial Brasileira, Secex/MDIC, December 2009).

Ranking	Firm	2008	2009	Var. % 2008/2009
3rd	Bünge Alimentos S/A	5,023.4	4,343.9	-13.53
5th	ADM do Brasil LTDA	2,769.5	2,592.9	6.81
6th	Cargill Agrícola S/A	2,205.2	2,335.6	5.91
8th	Sadia S/A	2,424.2	1,873.2	-22.73
10th	Louis Dreyfus Commodities Brasil	1,759.1	1,536.6	-12.65
11th	Brasil Foods S/A	0	1,505.7	–
14th	Amaggi Exportação e Importação LTDA	989.6	1,404.9	41.97
17th	Bertin S/A	1,204.7	1,165.3	-3.27
19th	Copersucar	897.3	1,062.2	10.24
20th	Seara Alimentos S/A	1,152.6	986.4	-14.4
22nd	Aracruz Papel e Celulose S/A	642.6	980.2	-21.73
23rd	Suzano Papel e Celulose S/A	1,256.4	921.4	-26.6
25th	JBS S/A	1,265.5	863.8	-31.74
29th	Cosan Indústria e Comércio S/A	478.7	756.0	57.94

a share of 17.1%. By 2009, their share went down to 7% and was surpassed by the increase in share of Africa, Middle East and China (Figure 9).

In 2008 China became the main destination of Brazil's agribusiness exports for the first time, overtaking the USA and the Netherlands. Compared to 2007, China purchased 69.7% more agribusiness products made in Brazil in 2008. The highest increase in imports of Brazilian agribusiness products took place in Venezuela, which bought 111.9% more in 2008 than in 2007.

The importance of agribusiness to the Brazilian economy is also to be seen in other figures. In 2008, Brazil's GDP totalled US$ 1.57 trillion, of which agribusiness corresponded to 26.5% (US$ 416.9 billion). Agriculture was responsible for US$ 293.9 billion dollars (70.5%), while livestock generated US$ 123.0 billion (29.5%) (Figure 10). Regarding the distribution of the agribusiness GDP in the agribusiness system, pre-farm activities generate around 6% of the total amount, on-farm activities 30%, and post-farm activities correspond to around 70%. As far as employment is concerned, 37% of employed people in the country were working in the agribusiness sector in 2005.

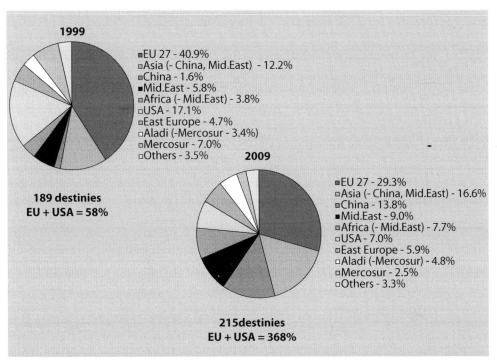

Figure 9. Brazilian agribusiness exports by destination (AgroStat Brasil/MAPA, data elaborated by GV Agro).

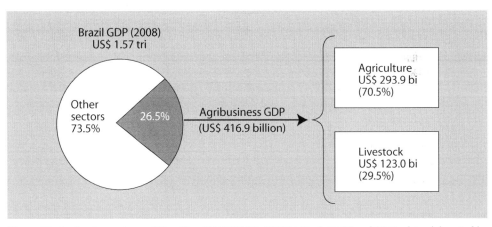

Figure 10. Agribusiness share of Brazilian GDP (IBGE, CEPEA/Esalq/USP and CNA, data elaborated by GV Agro).

1.6 The scenario for TNC investments: trends and global issues

The world has been facing major changes which are shaping the food and bioenergy markets. Those changes are related to four dimensions: the political-legal environment; the economic and natural environment; the socio-cultural environment; and the technological environment.

1.6.1 The political-legal environment for investments in Brazil

Investors must understand the role of protectionism. The resistant protectionism of countries is the predominant theme in the international media. There is an expectation that the distortions in world trade due to *protectionism* tend to be reduced gradually, because of the following factors:

- Increasing disputes in the WTO – World Trade Organization – as well as rounds of negotiations, which aim to reduce protectionism.
- Pressure from consumers in the developed world who are concerned about the social problems generated in the developing world for their subsidies based on production, stimulating overproduction and low prices for food ('anti-subsidies movements').
- Expansion and presence of multinational companies, which now dominate world trade. These companies will invest in the countries with greatest potential for exports. To these companies, it is not where the product is made that is important, but the return for the shareholders.
- The ageing of European, American and Canadian farmers leaves a new generation that is less interested in agriculture and did not go hungry during the world wars (main motivation for the policies on food security). This generation may be more liberalising than the previous one.
- Bioenergy. As the world moves towards renewable energy, the entry of products from agribusiness, but transformed into energy, will take on a new dimension, in that we have new allies such as environmentalists and entities of the third sector, provided that production is sustainable. Bioenergy represents a revolution in worldwide agriculture.

Moreover, there are fears about the *growth of terrorism and unstable regions*, which may generate concern about the supply of food and movements, stimulating a worldwide aversion to first generation biofuels from countries where the production systems are not 100% sustainable.

For all that, the expectation for the coming years is a reduction in tariff barriers and an increase in non-tariff barriers (health, social and environmental), e.g. for chicken (nitrofuran, avian influenza, Newcastle), citric pulp (dioxin), soybean (carboxin) and beef (mad cow, aphthous ulcer, deforestation of the Amazon). These *regulations accumulate and become increasingly complex*, requiring good planning and quick responses by countries in order to avoid these barriers. Increasingly it will become a battle of communication. It is in this scenario that TNC investments will operate.

Production with *sustainability* (environmental, economic and social), taking care of people, the preservation of the planet and profit maintenance, is the only way to produce. And therein lies the focus of protectionism. Environmental and labour standards will be stricter, leading to more stringent international missions looking at the patterns of production conducted mainly in the developing world, demanding quality and certification, as well as the *proliferation* of NGOs, aiming for the widest array of interests.

From this perspective, we have the recent ratification of the Kyoto Protocol, a pioneering plan to reduce emissions of developed countries, with different goals, and possible future goals for the emerging countries (China, India, Brazil). That, and the climate of political instability in the Middle East and the high dependence on the OPEC countries for the supply of oil, explains the current fever for biofuels (ethanol and biodiesel). The recent disclosure of the devastating effects of global warming will accelerate this process further.

1.6.2 The economic and natural environment for investments in Brazil

In the last 30 years, the world population has grown by a billion people, most of them in developing countries. And, despite the present lower birth rates, the world population will continue on its *growth trajectory*. The world will have 7.2 billion consumers in 2015, following an annual growth rate of 1.1% (FAO and OECD, 2007). And this increase in people feeds the global demand for grains and animal protein. In that context, Asia, and especially China and India, stands out as a great market with growing populations and income.

However, the exponential growth occurring mainly in Asian countries is not equitable. In this century, therefore, one of the greatest challenges to be faced by humanity is social inequality. Some of the solutions include the universalisation of health services and education, or, better still, the *distribution of income* and the inclusion of more people for a more sustainable life.

This situation of inequality is present in most developing countries, with an expected increase in *migration in the next few years*, affecting the labour supply in the world. These flows have already represented major problems for Europe (mainly the African population) and the USA (mainly the Hispanic population).

Increasingly multinational companies continue their process of *transfer of manufacturing*. They will produce where it is most interesting for their shareholders. The process of *globalisation of branding* is here to stay. The multinationalisation of companies makes the brands available, promotes competition and favours the consumer in the first place.

The *globalisation of the financial system* makes the resources flow more quickly between countries, which on the one hand brings instability, and on the other creates a necessary transparency and predictability in policies.

Increasingly, the economic environment will witness the growth of *partnerships and strategic alliances*; so that companies can focus their target activities and have strategic partners in production, supply and distribution. Interconnected systems, globalised, small, medium and large partners will be the agribusiness world.

Outsourcing of production activities is also a trend. Today, production assets are no longer needed for producing. The sharing of production capacity, even among competitors, has become popular. The processing industries of agribusiness challenge the focus on core competence and use of contracts for the supply of agricultural raw materials.

Furthermore, there are daily news items about *mergers and acquisitions* in the markets, involving a smaller number of larger companies, a move that is also happening in agribusiness. The greater competition, in most of the markets, bringing with it lower margins in the production systems, requiring a continuous process of cost reduction and re-evaluation of production processes.

Looking at bioenergy, it is noticeable that the scenario of continuing high oil prices, as well as the approval of environmental agreements to reduce carbon emissions, keeps the world's attention focused on *renewable energy*. This can already be seen by the fact that the major oil companies are changing their strategic focus, promoting themselves as energy businesses and including alternative energy sources in their strategic planning (ethanol, cellulosic ethanol, biodiesel and bio-esters), and thereby becoming more involved in sustainability.

Among the renewable energies, *biofuels* are at a higher stage of development and may compete with those derived from oil. Moreover, they are a good alternative to the crisis in agriculture, diversifying markets and improving the profitability of producers of grains, sugarcane and other oilseeds.

In general, the economic scenario is favourable to TNC foreign direct investments into agriculture.

1.6.3 The socio-cultural environment for investments in Brazil

A reality in the international context, which is not only characteristic of developed countries, is the *ageing population*. There is not only the demand for specific products for this segment of consumers – imagine the impact on national welfare systems – but also the need to consolidate the movement of demanding consumers, with high purchasing power and the willingness to pay for convenience.

As the population ages, another social problem causing global health care expenditure to rise is *obesity*. A serious consequence of living in cities, working long hours, eating an unbalanced diet and leading a sedentary lifestyle, is overweight and all the negative consequences to human health that come with it. This has stimulated the development and growth of the market for functional, organic and *diet/low-fat* food.

There is an increasing concern among countries about *food safety*. And, in this process, it is the final consumer who, regardless of nationality, will boycott 'suspect' products. After all, it is the obligation of the company, in response to the rights of the consumer, to adopt detection, prevention and elimination (traceability) systems for finding contaminants in food, eliminating risks and processing products.

It is also worth noting *the globalisation of lifestyles*, arising mainly from the communications revolution. Increasingly, societies are communicating, learning, and copying, and have styles in common. Tribes in all parts of the world are formed with similar behaviours; tribes that can compare products worldwide, that have enough information and want to be part of corporate life, interacting digitally with each other and being respected.

However, the social environmental appeal is more fashionable these days. The *environmental movement*, or green wave, along with the *organic food movement*, has put enormous pressure on the production chains.

Another interesting development is the *fair trade* movement. Formed by non-governmental organisations (third sector), they are responsible for buying and distributing the products of small growers from poor countries at premium prices, enabling them to develop their local communities.

In the last ten years a movement involving the broader concept of *sustainability*, a subjective and complex focus of which the environment is part, has attracted attention. For many consumers, it is not enough for a product to be 'green' (the mode of production must also be sustainable.

This concept was 'called' the 3Ps of sustainability: People, Profit, Planet: the concern that the organisations must have for the people involved directly and indirectly with the business, the profit that ensures the continuity of the attractiveness of the investment and, ultimately, the concern for the environment.

No attempt to differentiate the product or service through the appeals listed above will be valid if the company does not standardise its offer, following the precepts of a widely recognised *certification process*. This enables companies to monitor their production processes, ensuring the supply of products with certain attributes, and, concurrently, allowing the consumer to be able to distinguish the desired product from the 'copy'.

1.6.4 The technological environment for investments in Brazil

In the food industry, innovation in the area of creating more nutritious, more practical and more secure foods is demanded by the consumer. The TNC has a very important role to play here.

Regardless of where the innovation has to occur, mainly in the area of increased efficiency in the production of food, fibre and bioenergy, with lower consumption of non-renewable natural resources, the occurrence of environmental phenomena is here to show the way:
• The emergence of pests and diseases, which affect the main production chains on a global scale, is increasingly common.
• There is pressure for society to use less water, so water availability becomes a source of competitive advantage.
• Deforestation, mining, inappropriate agricultural practices and global warming causing desertification in the world, generating a loss of millions of hectares in agricultural land.
• The impact of climatic disasters changing the global configuration of production chains (hurricanes, tornados, earthquakes, floods, etc.).

It is necessary to develop new varieties of raw materials for biofuels, resistant to pests and diseases, and more suited to arid regions. On this point, genetic modification is an irreversible process, which is being expanded to several other crops.

In this way, it is also necessary to start examining the possibilities of so-called 'ecoefficiency'. In other words, the valorisation of clean technologies, rational use of resources, institutional adjustment, recovery of by-products, minimisation of environmental impacts and use of biodegradable materials for packaging. And in this analysis it is true that improvements in energy efficiency, allied to the use of renewable energy, tend to generate a much greater (and cheaper) reduction in CO_2.

Therefore, two things must be combined: energy efficiency and renewable energy. This is the reason for the global fever for the development of second-generation biofuels (technology for conversion of cellulose biomass, production of ethanol using enzymes), so as to have a world supply of *flex-fuel* cars and trucks (ethanol, petrol, natural gas, *diesel* and biodiesel), *hybrid* cars (electric batteries alongside fuel combustion) and also *hydrogen cells*.

Combining energy efficiency with renewable energy, and generating certified emission reductions which can be traded between countries, is even better. Today, the market for carbon credits modifies the way in which we 'consume' energy-generating reductions in CO_2 emission. Reductions in CO_2 emission, when audited by independent companies accredited in the Executive Board of the UNFCCC (United Nations Framework Convention on Climate Change), can generate certificates that are marketable in the world, enabling wealthier countries to meet their Kyoto targets for CO_2 reduction.

In order to make this a reality, it is necessary to create incentives for the entrepreneurs around the world to continue their work in generating innovations in a secure environment. The most correct way for this to happen is *international patenting*, or some form of international registration which respects the intellectual property rights and guarantees the entrepreneurs the claims on the research.

2. The sugar-energy value chain in Brazil

2.1 The sugar-energy supply chain

'Sugarcane is the world's leading feedstock for energy production'
(John Melo, CEO of Amyris)[1]

The Sugarcane Agribusiness System (AGS) is complex: mill production depends on the supply of sugarcane and capital goods. The main products (ethanol, sugar and energy) are sold to fuel distributors, the food industry, wholesalers, retailers, exporters and electric energy distributors. The by-products are destined for industry, wholesalers and retailers from other sectors, such as those of orange juice and animal feed. Today, the mills use the residues, such as vinasse and cake filter, as biofertilisers (Figure 11).

The sugarcane business is made up of many links: the production of sugarcane; the processing of sugar, ethanol and derivate products; the services for research, technical assistance and financing; transportation; commercialisation; and export. All of these links build a network around the mills as shown in Figure 12.

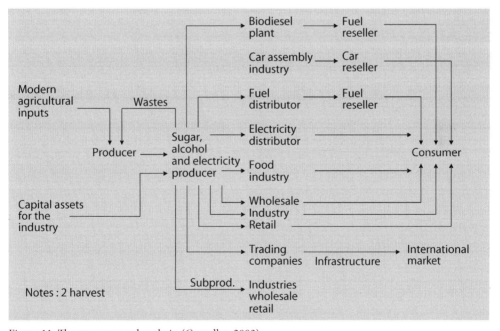

Figure 11. The sugarcane value chain (Carvalho, 2003).

---

[1] Speech done at the Ethanol Summit, São Paulo, Brazil, 2 May 2009.

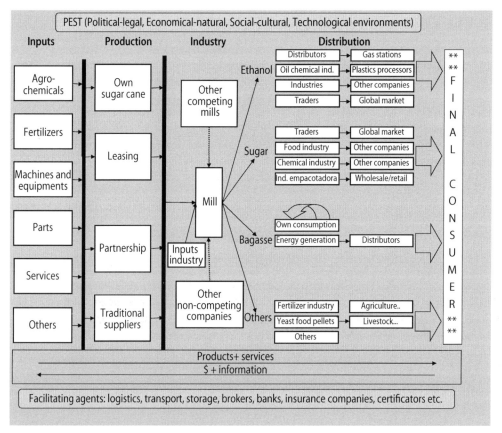

Figure 12. The network of a sugarcane mill.

There are different institutional arrangements that regulate the transaction of sugarcane from the farm to the industrial plants, such as spot markets, contracts or vertical integration. As Williamson (1985: 68) points out, 'the TCE states that this (contractual) diversity can be explained, mostly, by the basic differences on the transaction attributes'. Such dimensions are important due to the bounded rationality and opportunistic behaviour assumptions, and the ruling institutional environment.

The transaction asset specificity refers to how specific the investment (asset) is for the activity and how costly the reallocation of this asset is for another use (Williamson, 1985), or, in that case, how costly the devaluation of the asset is (Klein *et al.*, 1978). According to the Neves *et al.* (1998) and Moraes (2000) view of the sugarcane chain, the specificity is an important component of the analysis, since there are known specificities:

- *Location specificity*. Sugarcane is an input that cannot be transported over long distances. Ideally, the sugarcane transportation perimeter should not exceed 50 kilometres, due to transportation costs.

- *Temporal specificity* is due to excess supply that occurs during certain periods of the year. Sugarcane must be available for processing for eight months of the year. Another factor is the short shelf life of sugarcane after the harvest (48 hours).
- *Physical specificity* is great due to the industry (equipments) and because sugarcane is a long-term crop (5 year cycle), since the investments must be met in six exploratory years or five harvests.

Finally, the unpredictability factor, according to Williamson (1996), relates to the inability to foresee events that might influence the transaction. The unpredictability factor associated with possible opportunism inflicts extra costs on those transactions via the market, leading to the search for alternative ways of governance (Zylbersztajn, 1995).

The supply of sugarcane accounts for almost 70% of the mill production costs and the sugarcane transaction to mills is complex. The governance forms that exist between the producer and the sugar and ethanol plant are the contracts, the vertical integration and the spot market. Vertical integration can be observed when the sugarcane is grown in areas owned by the industrial plant. Leasing is the next option for de-verticalising. The following options, less integrated, are partnership, traditional supply contract and spot market contracts. There is a trend towards contractual relations in this transaction.

Vertical integration has been the dominant pattern of governance in the industry, with the extra feedstock needed for production being purchased from third parties. NIPE/UNICAMP (2005) estimates that 65% of the area cultivated with sugarcane is either owned or leased by mills, while 35% belongs to independent producers (around 70,000) – mostly under some form of contract.

It is important to note that the sugar industrial plants have risen in a vertical integration from the sugarcane growers. Thus, the dominant pattern of governance of this transaction until recently has been vertical integration, and the extra raw material needed for production is purchased from a third party.

Table 8 summarises the forms of governance of sugarcane transactions to the industry, with the associated risks and benefits. The general increase in costs, and the reduction in profit margins, drives the search for greater efficiency in the sugarcane production. If this were to be done inside or outside the mill, it would depend on the particularities of each mill and its skills in relationship management.

Table 9 gives an idea of the process of determining the governance structure, considering all the factors of local competition, land profitability, asset specificities, etc.

Table 8. Types of purchasing strategies used by the mills.

Types	How it is done
Vertical integration (own areas)	The mill owns the production asset (land). The mill manages all the agricultural activity.
Leasing	The mill leases the areas (minimum 5 year contract). The mill manages all the agricultural activity. The mill pays the landowner with a fixed % of sugarcane production.
Partnership contract 1	The mill transfers the land that is in its possession (owned or rented) to a partner. The partner manages all the agricultural activity. The mill receives a small share (20%) of the sugarcane revenues.
Partnership contract 2	The partner (landowner) prepares the land and manages the crop. The mill does the sowing, cutting, loading and transporting (CCT).
Partnership contract 3	The partner (landowner) transfers his land to a mill. The mill manages all the agricultural activity. The partner receives a small share (20%) of the sugarcane revenues.
Partner supplier contract	The partner (landowner) produces the sugarcane. The mill does the cutting, loading and transporting (CCT).
Traditional supplier contract	The supplier produces and delivers the sugarcane.
Spot sugarcane (gate)	The supplier produces and delivers the sugarcane. Producers without purchasing contract. Take advantage of better prices of delivery, in spite of risks.

Advantages	Risks
Total control of supplies.	High asset immobilisation.
Production planning and flexibility in decision-making (varieties, harvest period, industrial and transport optimisation).	Profitability depends on the land costs (hectare price).
Appropriation of agriculture activity results.	Health and climate risks.
	Occupied areas.
Financial balance between plant and production, due to price transferring.	Operational and administrative costs.
No need to buy the area.	Leasing contract hold-up.
Total control of supplies, which allows planning.	Better viability of competing cultures.
Profits from the local hectare cost.	Vulnerability of contract re-negotiations.
	Climate risks, land occupations and administrative costs.
No need to buy the area.	Leasing contract hold-up.
Sharing of climate risks.	Vulnerability of contract re-negotiations.
Greater control over selecting the producers.	Supply safety.
	Planning difficulty.
No need to buy the area.	The mill does not have total control of the crop management.
Fewer investments in soil preparation and crop care.	
Sharing of economic risks.	Weaker relationship and less trust.
Sharing of climate risks.	The landowner might switch to other crops.
Increased production in the areas surrounding the mill.	
No need to buy the area.	The same for the leasing operation.
Total control of supplies, which allows planning.	
Remuneration of the landowner linked to the levels of sugarcane production (variable).	
Same advantages as in partnership.	Same as partnership 2.
Mill structure optimisation.	Risk of competing with areas leased by the mill.
No need to buy the area.	Supply safety.
Administrative and operational costs are reduced.	Planning difficulty.
All production risks are on the supplier side.	Loss of advantages of vertical integration listed above.
Can negotiate with producers with lack of options, setting the prices low.	Risks of supply hold-up.
	Production planning difficulties.

Table 9. Factors in the decision for supply governance (sugarcane purchasing).

Factor	Decision
Presence of other mills in the same area	Creates competition for the sugarcane production among mills and, thus, more sales options for the producers.
	Then the mills need to seek better ways to recover their control of production.
	On the other hand, other mills in the same area might optimise the transportation, increasing their perimeter of action.
Hectare cost and land price trends	It is a variable that demands a high performance from the mill.
	If the cost is high, the tendency is to use someone else's land.
	The asset immobilisation and perspectives of land value should be considered.
	The presence of another mill in the region should increase the value of leasing over time.
Presence of qualified suppliers	Availability of producers with technical skills to manage the sugarcane.
	May need technical assistance.
	Qualified suppliers might make the mills opt for third party production.
Mill capacity	Depending on the mill infrastructure.
Agricultural mill capacity	Depending on the internal capacity for sugarcane production.
Financial mill capacity	Depending on the mill cash flow and availability of capital.
	Conditions for acquiring credit and warranties.
Land competition between other existing crops	The more different the crops, the more control is needed.
	Competition with grains.
	Longevity of sugarcane crops limits the ability to change to other crops.
Price trends and volume produced.	Unstable prices require greater control.
	Hedging operations – own sugarcane price is linked to production costs, while third party sugarcane price is linked to sugar and ethanol prices (Consecana).
Risks of sugarcane production	Climate and health risks.
	Supply safety.
Coordination capability (contracts)	Capability for contracts and relationships management.
	Low value (margins) in the sugarcane business.
	Production seasonality.
Advantages of integrated operations	Fiscal advantages.
	Economy of scale and scope.
Cultural aspects and capability for association	Political and social pressures related to labour conditions and environment.
	Capability for association of producers because of cultural aspects.
Logistics	Time efficiency between harvest and transport.
	Better operational rationality to the perimeter of action.
Inputs costs	Access and proximity of the inputs' suppliers and capability of negotiation.

In this sense, a theoretical contribution is made to the sustainable production of sugarcane on new agricultural frontiers. Figure 13 details this 3-step process.

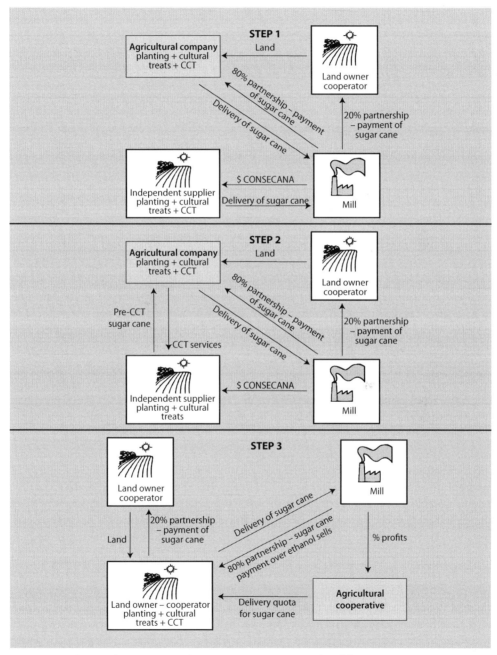

Figure 13. Evolution of sugarcane supply governance in new agricultural frontiers.

In Step 1, the supply is ensured by the mill's own Agricultural Company and by independent suppliers of cane (if there are any). The Agricultural Company will establish a major partnership contract with the small integrated producer, since he is the one who has possession of the land. In this agreement, the Agricultural Company gets 80% of the revenue, because it is responsible for all stages of production (planting, crop treatments and harvest/CCT). The small producer gets 20% and over time he will learn how to produce and capitalise himself. At the same time, an association or cooperative of producers will have a minority stake (shareholder) in the mill to be capitalised as well.

In Step 2, the cane supply has new suppliers. These actors are the small integrated producers that were trained to do the planting and the crop treatments for the cane. However, they let the Agricultural Company do the harvesting (CCT), because they are still financially unable to do so even through a labour or machinery pool. The small producer will receive a price per tonne of cane (valued by Consecana) minus the cost of the CCT. There are still some small producers who resist the organisational change and prefer to work in the majority partnership system described above.

Finally in Step 3, a cane supplier takes care not only of the planting and crop treatments, but also of the harvest (CCT). Accordingly, the landowner, the small integrated producer, can also be the cane supplier. Therefore, some entrepreneurs, who do not grow exclusively on their own land, also take care of the crops of other small producers who are less able, in a system of agricultural partnership. In any case, through an association or cooperative, producers can have a labour pool to do the CCT and thus share the resources. This cooperative should also be able to buy machines for mechanical harvesting, as it receives dividends from the mills, and start to dictate the agricultural planning.

2.2 The sugar-ethanol evolution in Brazil

2.2.1 The sugar evolution in Brazil

The production of sugar was an important instrument for the settlement of the Portuguese colony in the Americas. During the first 30 years after the Europeans first arrived in the territory which is now known as Brazil, the most important economic activity in Brazil was the exploitation of 'pau-brasil', the wood after which the country is named. For that purpose, there was no need for settlement or the establishment of a political-administrative system.

In 1530, taking advantage of their experience in producing sugar from cane in other colonies, the appropriate tropical weather and the valorisation of sugar on the European market, the plant was introduced in Brazil and the sugarcane plantation became the basis of the settlement, around which the first administrative structure and the regular trade between colony and metropolis were created. Brazil soon became the world's largest sugar producer. Just over one century later, sugar prices in Europe dropped significantly as the

Dutch expanded production in their tropical colonies. At the end of the 16th century the product lost importance with the advent of mining, especially gold, but it would continue to be important throughout the country's history.

Currently, sugar is recognised as the basic source of energy for our metabolism and the whole food and drink industry depends on it. Because of its importance, almost all countries produce sugar, either from sugarcane or sugar beet. The largest producers are Brazil, India, China, the USA, Australia and Thailand. According to data from the United States Department of Agriculture (USDA), Brazil's production reached 34.6 thousand tonnes in the 2009/2010 crop year, a record that represents 46.5% of all sugar produced in the world. The country is just as significant in terms of world trade. Out of the total production, Brazil exported 20,250 thousand tonnes, or 45% of the world's exports (USDA, 2010).

Brazilian sugar production has experienced almost continuous growth in the last 20 years. First, with the crisis of the Alcohol National Program (ProAlcool), the liberalisation of the sector and the high sugar prices in the international market, many of the mills directed their production mix in favour of sugar at the beginning of the 1990s. More recently, the investments in the production of ethanol, which followed the launch of the flex-fuel automobiles and the growth of the international market, are also benefiting the production of sugar, since the vast majority of mills produce both ethanol and sugar. Figure 14 shows the evolution of Brazilian sugar production, which has grown 577% in 40 years. India, which is usually a major exporter, became Brazil's largest buyer in 2009 after consecutive years of climate conditions that affected the production of the world's largest sugar consumer. On the other hand, Russia's imports from Brazil fell to about half the volume bought in 2008 as a result of the country's economic setbacks. As a result, in 2009 India imported 18% of all sugar exported from Brazil, followed by Russia (11%), the United Arab Emirates (7%), Bangladesh and Nigeria (5%) and Saudi Arabia (4%).

Sugar consumption is generally influenced by the price of sugar and alternative sweeteners; the existence and availability of stocks; the preferences of consumers; and public policies. Nevertheless, per capita income and population growth are the most important drivers of demand.

Worldwide sugar consumption has increased by almost 10% in the last 10 years, reaching 156,650 thousand tonnes in 2009 (USDA, 2010). India is the largest consumer market with a share of 15%, followed by China (12%), Brazil (9%) and the United States (6%).

In Brazil, the recent increase in income due to economic growth and the government's income distribution policies has led to a growing demand for industrialised products. The food and drink industry is the largest sugar consumer and has consequently leveraged sugar consumption. USDA market year data show that the Brazilian internal market consumed 12,250 thousand tonnes of sugar in 2009/2010, a 4.4% increase compared to the previous

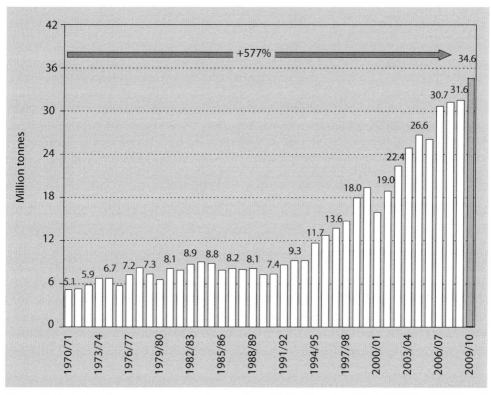

Figure 14. Evolution of Brazilian sugar production from 1970/1971 to 2009/2010 (MAPA, CONAB and USDA, data compiled by GV Agro).

year. Retail is responsible for 40% of total consumption while industry is responsible for 60% of internal demand.

Within industry, 20% of total demand is used for producing soft drinks and 10% for producing sweets and chocolate. Considering that per capita chocolate consumption in Brazil is 15 times lower than in Sweden and 10 times lower than in the USA, and that soft drink per capita consumption in the USA is 4 times higher, there is still a lot of room for the growth of sugar consumption through industrialised products.

USDA data showed a slight increase of 1% in sugar consumption worldwide in 2009/2010 despite the international financial crisis. This represents a timid sign of recovery after the combination of consecutive years of low international sugar prices; climate restrictions suffered by some of the main producers, such as India, Thailand and Australia; and the end of subsidies for production and exports in the EU, after a deal at the WTO, which resulted in a 10.7% reduction in global production in 2008/2009 compared to the previous year (BRADESCO, 2009 and USDA, 2010).

Brazilian sugar producers have been benefiting from this scenario in terms of price and market share. The country's exports increased 23% in 2009/2010, partly filling in the gaps left by the decrease in external sales from the EU and supplying the now debt-ridden Indian market. Therefore, those mills which produce both sugar and ethanol have been able to compensate to some extent for the low prices of ethanol with the gains from sugar sales, as well as generating cash flow to be able to keep on operating despite the amount of debt accumulated in the previous year and a half.

Table 10 shows the evolution of Brazilian sugar exports from 1999 to the first two months of 2010.

2.2.2 The ethanol evolution in Brazil

The industrial production of ethanol fuel in Brazil started in the 1930s, stimulated by the first government incentives (for information about the use of ethanol as motor fuel, see Box 1). A federal law from 1931 mandated a 5% ethanol mix in all imported petrol. In the same year, all public service cars had to run with a 10% ethanol mix, and in 1938 the 5% mix also became mandatory for petrol produced in the country. However, it was not until the 1973 oil crises that sugarcane became an important component of Brazil's energy matrix. At that time, 77% of the oil consumed in the country came from abroad. Oil imports shot up from US$ 760 million to US$ 2.9 billion within one year (Carvalho, 2004).

Table 10. Evolution of Brazilian sugar exports from 1999 to 2010 (Secex, data compiled by GV Agro).

Year	US$ million (FOB)	Million tonnes	Average prices
1999	1,911	12.100	157.91
2000	1,199	6.502	184.41
2001	2,276	11.168	203.92
2002	2,090	13.344	156.65
2003	2,140	12.914	165.71
2004	2,640	15.764	167.49
2005	3,919	18.147	215.95
2006	6,167	18.870	326.81
2007	5,100	19.359	263.47
2008	5,539	19.721	280.87
2009	8,376	24.294	344.85
2010	1,426	3.176	449.06
2009 (Jan-Feb)	1,008	3.371	298.95
2010 (Jan-Feb)	1,426	3.176	449.06

Box 1. Ethanol.

Ethanol, also known as ethyl alcohol, can be produced by the fermentation of sugarcane juice and molasses. It has been used in various forms for thousands of years, and has recently emerged as a leading fuel for combustion engines. Since March 2008, ethanol represents more than 50% of Brazil's overall petrol consumption. Brazil produces two types of ethanol: hydrous, which contains about 5.6% water content by volume; and anhydrous, which is virtually water-free. Hydrous ethanol is used to power vehicles equipped with pure ethanol or Flex-Fuel engines, while anhydrous ethanol is mixed with petrol before it reaches the pump. Several countries are now blending anhydrous ethanol with petrol to reduce petroleum consumption, boost the octane rating and provide motorists with a less-polluting fuel. Brazil is a pioneer in using ethanol as a motor vehicle fuel. The country began using ethanol in automobiles as early as the 1920s, but the industry gained significant momentum in the 1970s with the introduction of ProAlcool, a trailblazing federal program created in response to global oil crises. ProAlcool succeeded in making ethanol an integral part of Brazil's energy matrix, but the programme faced numerous challenges, particularly in the late 1980s when oil prices tumbled and sugar prices were high. Ethanol use blossomed again in Brazil because of sky-high petrol prices, environmental concerns and the introduction in 2003 of Flex-Fuel vehicles (FFVs) that can run on ethanol, straight petrol or any mixture of the two (UNICA, 2007).

Aiming to reduce the negative impacts of oil prices in the trade balance, the Brazilian government launched the Alcohol National Programme (ProAlcool) in 1975, starting a series of large investments in the development of ethanol-burning engines and stimulating the production of sugarcane and its products through tax cuts, price controls, strategic stocks, special lines of credits and mandatory blending and distribution. Between 1975 and 1978, the demand for anhydrous ethanol (used in non-ethanol engines for blending purposes) jumped from 1.1% to 9% of total fuel consumption. In 1979 the first ethanol engine car was launched on the market. In 1986, the percentage of ethanol cars as a proportion of the sales of new cars reached 95%.

However, in the late 1980s and early 1990s, oil prices fell; the Brazilian government promoted the deregulation of the sector, ending subsidies and shrinking credit; and mills responded to high sugar prices by shifting industrial production in favour of sugar. Soon, ethanol prices rose to the same level as petrol, the strategic stocks were sucked dry and the drivers of ethanol cars found themselves literally out of fuel, which was a major blow for the image of the milling sector.

The launch of the flex-fuel cars in May 2003 reinstated trust in ethanol among consumers and car makers. With this type of car, drivers could just fill up their tanks with petrol in the event of a shortage in the supply of ethanol. In 2009, a record 2,993 million cars were sold in Brazil, leaving behind Spain and France and becoming the sixth largest producer. In that same year, 92.6% of the new cars sold in the country were flex-fuel (ANFAVEA, 2010).

Internal ethanol demand stagnated at between 11.5 and 13.0 billion litres from 1986 until 2007. In 2009, with the growth of the market for flex-fuel, demand reached 22.8 billion litres, i.e. 16.4 billion for flex fuel cars and 6.3 billion for the mandatory blending that ranges from 20% to 25%. In February 2008, ethanol consumption overtook petrol for the first time since the peak of ProAlcool in the second half of the 1980s. According to the National Petroleum Agency (ANP, 2008), in that month Brazilian drivers used 1.432 billion litres of ethanol versus 1.411 billion litres of petrol. With the gradual switch from petrol cars to flex-fuel cars, ethanol consumption keeps on increasing as long as prices are favourable.

Recently, concerns regarding global warming and the instability of oil prices have led a growing number of countries to add ethanol to their fuel matrix. With a few exceptions such as Brazil, ethanol production is heavily subsidised and most of the world's demand is produced domestically. Nevertheless, Brazilian ethanol has some important comparative advantages, especially in terms of costs and energy balance, which have helped it expand overseas sales (Table 11). Historically, the USA had been Brazil's ethanol major importer, until 2009 when the country's imports dropped from 1.54 billion litres in 2008 to 272.19 million litres, and were overtaken by the Netherlands (678.47 million litres), Jamaica (437.66 million litres), India (367.57 million litres), South Korea (313.71 million litres) and Japan (279.96 million litres).

Table 11. Brazilian ethanol exports 2000-2009 (UNICA, 2010 and Secex, 2009).

Year	Volume (millions of litres)	US$ FOB (in millions of dollars)	Average price (US$/m^3)
2000	227.3	34.8	153.07
2001	345.7	92.1	266.57
2002	789.2	169.2	214.35
2003	757.4	158.0	208.57
2004	2,408.3	497.7	206.68
2005	2,600.6	765.5	294.36
2006	3,416.6	1,604.7	469.69
2007	3,530.1	1,477.6	418.58
2008	5,118.7	2,390.1	466.94
2009	3,296.3	1,338.0	405.92

In order to meet the growing demand, production has more than doubled in a matter of years, rising from 11 billion litres in 2001/2002 to 26 billion litres in 2009/2010. Figure 15 shows the evolution of production since the 1970s.

The recent international financial crisis has deeply affected the ethanol market. Combined with the decrease in oil prices, it has caused a decrease in demand in the USA and the EU, and Brazil's total exports dropped by around 35.6% in 2009. In the domestic market, this is likely to reduce prices despite the continuously growing demand caused by the gradual substitution of petrol vehicles with flex-fuel cars.

2.3 Mapping and quantification of the sugar-energy supply chain in Brazil[2]

The objective of this section is to map and quantify the agro-industrial system of sugar-energy in Brazil. The product of this analysis is the estimated revenue of the different links

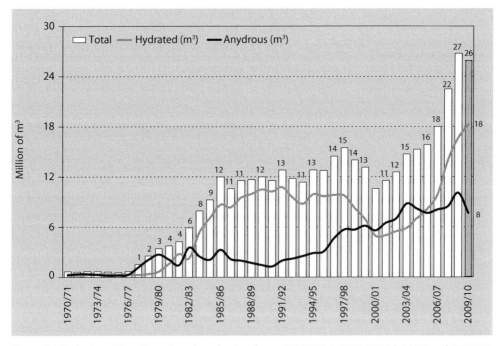

Figure 15. Evolution of Brazilian ethanol production from 1970/1971 to 2009/2010 (MAPA and CONAB, data elaborated by GV Agro).

[2] This section was based on the paper 'Mapping and Quantification of the Sugar-Energy Sector in Brazil' by Marcos Fava Neves, Vinicius Gustavo Trombin and Matheus Alberto Consoli, presented at the 2010 IAMA World Forum and Symposium. We would like to thank Professors Vinicius Trombin and Matheus Consoli for the permission to reproduce part of this paper.

of the productive chain in 2008 and the sectoral GDP. Besides financial transactions, other research focuses are the quantification of jobs and taxes generated in the sector. This research is part of a broader effort of the Brazilian Sugarcane Industry Association (UNICA), the largest organisation in Brazil representing sugar, ethanol, and bioelectricity producers. The association is working to increase knowledge about the sugar and ethanol industry and to convey to the public, in a concise and uniform way, the benefits of production and use of clean energy from renewable and sustainable agricultural origins.

The sugar-energy sector's GDP in 2008 was US$ 28,153.10 million, equivalent to almost 2% of the national GDP or almost the overall economic output produced in a country like Uruguay (US$ 32 billion). The sectoral GDP calculation was estimated by adding the sales of all final goods and services of this agro-industrial system.

As shown in Table 12, subtracting sales taxes, the amount is US$ 24,344.43 billion. The assumptions used to estimate taxes are presented in Section 2.7.1 'Taxes'.

Table 12. Estimates of the sector's gross domestic product based on the end products (Neves *et al.*, 2010; data generated by MARKESTRAT).

Product		Domestic market (million US$)		Export (million US$)	Total (million US$)	
		With taxes	Tax free	Tax exempt	With taxes	Tax free
Ethanol	Hydrated	11,114.50[a]	9,105.10	23.78	11,138.28	9,128.88
	Anhydrous	2,972.89[b]	2,250.88	2,366.33	5,339.22	4,617.21
	Non-energetic uses	438.78[c]	351.57	nd	438.78	351.57
Sugar		5,297.14[d]	4,455.83	5,482.96	10,780.10	9,938.79
Bioelectricity		389.63[e]	242.87	nd	389.63	242.87
Yeast		21.41	19.43	42.20	63.61	61.63
Carbon credits		nd	nd	3.48	3.48	3.48
Total		20,234.35	16,425.68	7,918.75	28,153.10	24,344.43

[a] Sales by gas stations, taking into account the formal and informal markets.
[b] Sales by the ethanol plants to ethanol wholesale distributors, taking into account the formal and informal markets.
[c] Sales by ethanol plants to the beverage and cosmetics industries.
[d] Sales by sugar mills to the food industry plus the sales by retailers to final consumers.
[e] Sales by the sugarcane mills and ethanol plants at energy auctions.

Figure 16 presents the sugar-energy agro-industrial system, and the values below each link indicate its gross sales in this productive chain in 2008. Total gross revenue of the sugar-energy sector (Gross Production) was about US$ 86,833.00 million. This value represents the sum of all estimated sales made by every link of the agro-industrial system to the sugar-energy chain and the financial transactions of the facilitating agents. Figure 18 presents the gross revenue of each link in the productive chain.

2.4 Pre-farm statistics

2.4.1 Agricultural inputs

The agricultural inputs industry sold about US$ 9,252.42 billion to the sugar-energy sector in 2008; this includes pesticides sales by agricultural cooperatives and dealers of US$ 477.54 million. Figure 17 summarises all the agricultural input revenues, which are detailed in the following text.

The sugar-energy sector accounted for 14% of the total agricultural fertiliser sales in Brazil in 2008, totalling about US$ 2,259.09 million (3,140 thousand tonnes). This input is essential for sugarcane production in Brazil. Thus, sugarcane is the third largest Brazilian fertiliser market, behind only soybean and maize. The increase in sugarcane production in recent years has caused an incremental rise in fertiliser demand, even with the unfavourable exchange ratio. While, in 2007, 19.8 tonnes of sugarcane were needed to buy a tonne of fertiliser, in 2008 the volume soared to 36.3 tonnes of sugarcane. This happened due to the rising price of fertilisers and the reduction in sugarcane prices. Sales of lime to sugarcane plantations in 2008 were estimated at US$ 50.56 million, corresponding to 2.999 million tonnes or 14% of national consumption.

In 2008, the pesticides industry had revenues of US$ 768.44 million with the sugar-energy sector, representing 9.5% of total sales in the country. The cooperatives were responsible for 61% of pesticide sales to the sector, and dealers represented 2%, together grossing more than US$ 477.54 million. Direct sales accounted for 37%. Of the amount spent on pesticides by farmers in the production of cane sugar, 73.5% went on herbicides, 22.8% on insecticides, and 3.7% on fungicides. Sugarcane production stands out among the 3 crops that consume most pesticides in the country, highlighting the importance of the sector.

About 3,970 tractors were sold to the sugar-energy sector in 2008, generating revenues of US$ 320.87 million. This amount represents 9% of total tractors sales in the country. When it comes to the category of tractors over 200 hp, the sugar-energy sector was responsible for buying 47% of the total. Revenue from agriculture implements was about US$ 425.66 million. This segment includes ploughs, disc harrows, subsoilers, and self-propelled irrigation systems, among other items. The autoparts sector jointly with machinery maintenance services enjoyed revenues of about US$ 2,851.19 million, including parts and labour to

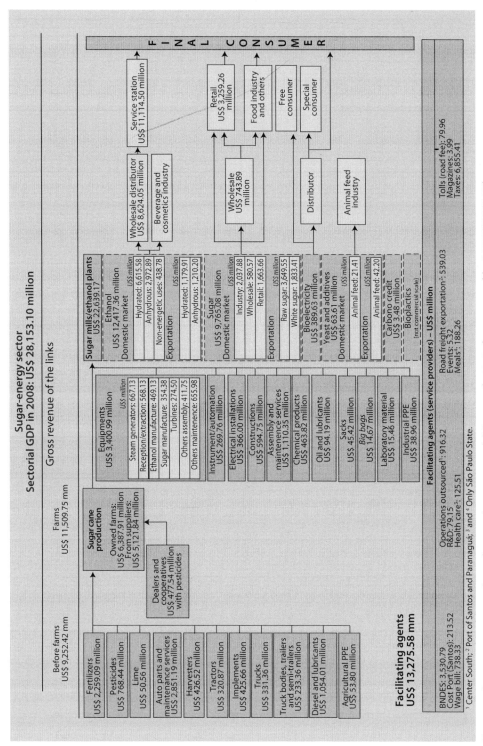

Figure 16. Sugar-energy agro-industrial system (gross revenue) (Neves et al., 2010; data generated by MARKESTRAT).

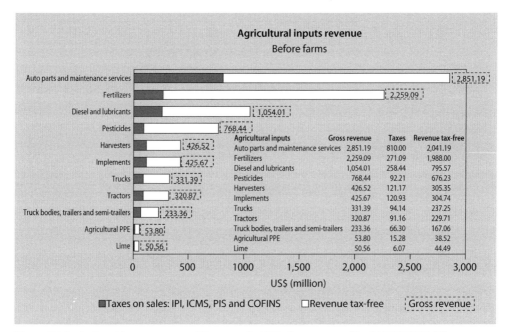

Figure 17. Agricultural inputs revenues (Neves *et al.*, 2010; data generated by MARKESTRAT).

keep nearly 144 thousand machines operational, each of which require approximately US$ 20,000 in maintenance per year.

The sugar-energy sector acquired 22% of harvesters sold in 2008, accounting for a turnover of US$ 426.52 million. This constituted 981 units sold to the sector, representing a growth of 52% compared with 2007. The national fleet of cane harvesters almost doubled. In 2007 there were approximately 1,280 harvesters on the sugarcane plantations. The legal deadlines for eliminating the burning of sugarcane straw prior to harvesting was one of the reasons that led to this significant growth in sales (the Agri-Environmental Sugarcane Protocol). Other factors that contributed to this result are the possibility of reducing agricultural costs and the lack of skilled manpower in various production regions, especially in the expansion areas, where sugarcane is not yet a traditional crop. In addition, the Protocol stipulates that all burning must cease in São Paulo by 2014 in areas where mechanised harvesting is possible. In areas where mechanisation is currently not possible, the deadline is 2017.

The sales of heavy trucks, a category with total gross weight over 40 tonnes, were also driven by the growth of the sugar-energy sector. In addition to transporting ethanol, these trucks deal with 80% of the harvested sugarcane transportation from farm to plant or mill. It is estimated that in 2008, 1,962 trucks were sold to the sector, or the equivalent of 5% of this truck category's sales countrywide. The financial transactions were estimated to be in the order of US$ 331.36 million.

Food and fuel

In 2008, sales of truck bodies, trailers, and semi-trailers were estimated at US$ 233.36 million. In addition to the 488 truck bodies sold, the license plates of 4,856 trailers and semitrailers were registered, which accounted for about 9% of total sales of heavy machinery in Brazil, and an 11% increase over 2007.

The agricultural mechanised operations and the sugarcane transportation from farm to ethanol plant and sugar mills consumed about 1,036 million litres of diesel fuel and lubricants, equivalent to US$ 1,054.01 million. Also, in 2008 the investments in worker safety reached US$ 53.80 million for personal protective equipment (PPE).

2.5 On-farm statistics: sugarcane production

The 2008/2009 sugarcane harvest reached a record production of 568.96 million tonnes and a planted area of about 8.5 million hectares. The São Paulo State accounted for 68.6% of sugarcane crushing in the south-central region. In this region, Minas Gerais was the state with the largest increase in production in the past 5 years, with growth of 1.8%, followed by the State of Goias at 1.6%.

Sugarcane was responsible for revenues of around US$ 11,509.75 million. The yield of raw material was 143.25 kg of total recoverable sugar (ATR) per tonne of sugarcane, a decrease of 2% over the previous crop. The ATR average value was US$ 0.14, and the average sugarcane price was US$ 20.23 per tonne. In the 2008/09 season, as shown in Figure 18, sugarcane from suppliers accounted for approximately 44.5% of the industry demand (US$ 5,121.84 million) and 55.5% were harvested on the farms owned by the ethanol plants and sugar mills (US$ 6,387.91 million).

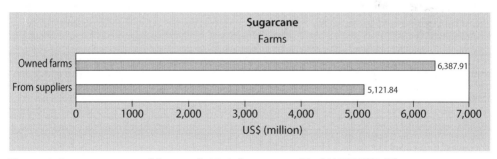

Figure 18. Sugarcane revenue (Neves *et al.*, 2010; data generated by MARKESTRAT).

2.6 Post-farm statistics

2.6.1 Equipment, industrial services, and supplies

The sugar-energy sector was responsible for the purchase of US$ 6,414.39 million in industrial inputs. This value is presented in detail below.

The industrial equipment sales and the billing of the companies that provide assembly services were estimated by considering the investments made in the 29 ethanol plants and sugar mills that started operations in 2008. It is known that these investments must have been started in 2006 and finished in 2008, so they do not represent the sector sales in this year. They do, however, represent an estimate of the financial transactions generated for the new unit installations, which began production in 2008. Of the 29 industrial units, we adopted the premise that 4 are sugar mills (3 have a milling capacity of 1.5 million tonnes of sugarcane per year, and 1 has a capacity of 3 million tonnes) and 25 are ethanol plants (15 with a milling capacity of 1.5 million tonnes, and 10 with a capacity of 3 million tonnes).

The average investment needed to assemble the industrial part of a sugar mill was estimated at US$ 85 per tonne of sugarcane milling capacity and for an ethanol plant approximately US$ 75 per tonne. Table 13 shows the proportion of the investment needed, and Table 14 details the investment in equipment.

In addition to investments related to the new units' installation, we also considered sales of equipment and services for the maintenance of industrial units, which is performed between harvests. We put the maintenance cost in the central-south Brazilian region at US$ 1.68 per tonne of sugarcane milled, i.e. 62.50% spent on equipment and 37.50% spent on services. In the north-east Brazilian region, this cost was US$ 2.08, i.e. 86.70% and 13.30% spent on equipment and services, respectively. Also considered were the automation and

Table 13. Proportion of investment among the items (prepared by MARKESTRAT from data provided by Procknor Engineering).

Item	% of the total investment
Equipment	60%
Electromechanical assembly	7%
Construction	13%
Electrical installations	8%
Instrumentation/automation	2%
Engineering services, thermal insulation, and painting	10%
Total	100%

Table 14. Proportion of investment of the equipment per kind of equipment (prepared by MARKESTRAT from data provided by Procknor Engineering).

Equipment	% of the investment in equipment	
	Sugar mill	Ethanol plant
Steam generators	25%	20%
Sugarcane reception, preparation, and extraction system	20%	25%
Ethanol manufacture	15%	30%
Sugar manufacture	15%	0%
Turbines/power generators	10%	10%
Others	15%	15%
Total	100%	100%

instrumentation projects sold in 2008 for the sugar-energy sector (which amounted to 41 projects more than those sold for the 29 new units).

Based on these assumptions, the revenue of the industrial equipment suppliers was estimated at approximately US$ 3,400.99 million. Sales of automation and instrumentation were US$ 269.76 million, and service providers of assembly and maintenance had revenues of approximately US$ 1,110.35 million. The construction sector generated about US$ 594.75 million, and electric installation US$ 366.00 million.

The sugar-energy sector generated revenues of US$ 463.82 million purchasing the products and specialty chemicals for ethanol and sugar production, including quicklime, polymers (auxiliary in the production of sugar and ethanol), yeast, water treatment, and ion-exchange resins, among other inputs.

The fuel and lubricating oil consumption for the industrial operation was 70 million litres, generating revenues of US$ 94.19 million. The cost of laboratory material was US$ 15.46 million. Costs for 50 kg sacks and 1,200 kg big bags amounted US$ 45.42 million and US$ 14.67 million in 2008, respectively. Industrial PPE enjoyed sales worth US$ 38.96 million. Figure 19 summarises the revenue generated by the industrial inputs.

2.6.2 Sugar mills and ethanol plants

The sugar mills and ethanol plants grossed US$ 22,639.17 million with all of the products, i.e. US$ 12,417.36 million with ethanol, US$ 9,765.08 with sugar, US$ 389.63 million with bioelectricity, and US$ 67.09 million with yeast, additives, and carbon credits. These

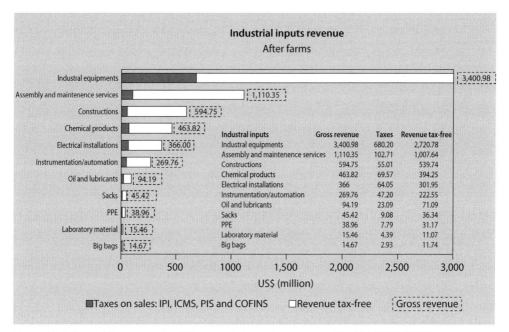

Figure 19. Industrial inputs revenue (Neves *et al.*, 2010; data generated by MARKESTRAT).

products represent, respectively, 55%, 43%, 1.7% and 0.3% of their sales. The products and their distribution channels are presented below.

2.6.3 Ethanol

The sugar mills and ethanol plants grossed US$ 12,417.36 million with ethanol in 2008, taking into account the domestic and international markets.

Exports generated revenues of US$ 2,366.33 million, i.e. US$ 1,179.91 million for hydrous ethanol and US$ 1,210.20 million for anhydrous ethanol. Exports of anhydrous ethanol were atypical in 2008. One reason for its growth was the increase in American demand, due to the decrease in crops because of the flooding in the main producing region of that country, in addition to the significant increase in the cost of oil, which exceeded US$ 100 per barrel during that year.

Brazilian ethanol exports totalled 5.12 billion litres. The main buyers were the United States (34%), the Netherlands (26%), Jamaica (8%), and El Salvador (7%). However, the total export volume is still small compared with total production, which already shows great potential for growth, with a 14-fold increase in volume and a 24-fold increase in currency since 2001. The most significant increase in volume occurred in 2004 (220%), when around 2.4 billion litres were exported.

The domestic market consumed 14.08 billion litres of hydrated ethanol in 2008 (formal and informal market), generating a turnover of US$ 6,615.58 million for the sugar mills and ethanol plants. Sales of hydrated ethanol have grown considerably in recent years. Compared with 2006, the increase was 87%. The main reason for this growth was the introduction of flex-fuel cars, which accounted for 90% of the light commercial vehicle production in Brazil in 2008.

The anhydrous ethanol in the internal market generated a turnover of US$ 2,972.89 million, with 6.48 billion litres in 2008 (formal and informal market). The major consumption of this product in Brazil is for blending with petrol, currently at the rate of 25%. The anhydrous ethanol consumption has decreased in recent years because of the rise in numbers of flex-fuel cars.

Ethanol for non-energy uses is destined mainly for the production of beverages, cosmetics, pharmaceuticals, and chemicals. According to data from the National Energy Balance, this consumption was 720 million litres in 2008, representing a turnover of US$ 438.78 million for the sugar mills and ethanol plants.

2.6.4 Wholesale distributors and service stations

Wholesale distributors earned US$ 8,624.05 million, and the service stations US$ 11,114.50 million. For the purposes of estimating revenues, it was considered that the informal market occurs between the ethanol plant and the service station, by-passing the distributor.

2.6.5 Sugar

The sugar mills earned US$ 9,765.08 million from sugar in 2008, counting sales to both the domestic and international markets. Exports generated revenues of US$ 5,482.96, i.e. 67% for raw sugar and 33% for white sugar. Of the 19.47 million tonnes shipped, 83% were produced in the Brazilian south-central region and 17% in the north-northeast region. About 50% of exports went to 5 countries, while the rest was exported to more than 100 other countries. Between 2000 and 2008, 25% of the Brazilian sugar exported was destined for Russia, the main international buyer market, followed by Nigeria, Egypt, Saudi Arabia, Algeria and others. The largest share of sugar production is destined for foreign markets. Production grew at rates much higher than the growth of Brazilian consumption, which has remained stable over the last 6 years at an average of 3% per year.

In the domestic market, the sugar mills' turnover was US$ 4,282.12 million from sugar. Of this total, sales to the food industry generated a turnover of more than US$ 2,037.88 million, revenues destined for the retail trade were US$ 1,663.66 million, and wholesale was US$ 580.57 million. Part of the sugar volume destined for industry is marketed through a wholesaler and is not sold directly by the sugar mill to industry. Generally, this transaction occurs with wholesalers specialised in small factories. These wholesalers, in addition to selling to the factories, sometimes pack the sugar and sell it for retail.

The main sugar-consuming industries are the producers of soft drinks (20%), confectionery and chocolates (10%), chemicals (10%), and milk (7%), with other industries accounting for 53%. The main type of sugar for fresh consumption is sugar crystal (61%), followed by refined sugar (36%), and refined grain and others (4%).

The central-south region sold 10.5 million tonnes, and the north-northeast region 1.02 million tonnes. 60% of central-south production was intended for industry, 28% for direct retail sales, and 12% for wholesale. The north-northeast production was 53% destined to direct retail sales, 25% to industry, and 22% to wholesale. Overall sales to industry totalled 6.59 million tonnes of sugar, direct sales to retail around 3.5 million tonnes, and wholesale sales were 1.49 million tonnes.

2.6.6 Wholesale and retail

The wholesale industry earned US$ 743.89 million with sugar, and the retail US$ 3,259.26 million.

2.6.7 Bioelectricity

The bioelectricity generated from sugarcane bagasse increasingly stands out as an important product of the industrial plants. Bagasse is the dry fibrous waste left after sugarcane is crushed. Sugarcane is composed of 1/3 juice, 1/3 bagasse and 1/3 straw. This bagasse is then burned in boilers generating steam and finally energy, used both for the mill's operations and also sold back to the electricity grid. In 2008, about 30 plants had negotiated 544 MW for annual sales over 15 years. This volume will generate annual revenues of US$ 389.63 million.

According to UNICA, in 2009, sugar and ethanol units had the potential to generate 4,034 megawatts of surplus energy, which represents around 3.58% of Brazilian energy needs. The projections with all the investments are that in 2015, around 11,500 megawatts can be produced, representing around 15% of the country's needs.

There are several benefits to this alternative, since it is low-impact and a company can also apply for carbon credits. Most of Brazilian energy comes from hydro-electrical facilities, however bioelectricity from sugarcane bagasse comes with the dry season, when major dams have lower levels of water, and is therefore very complementary to hydro-electrical power.

With the mechanisation of the harvesting process, a larger amount of green material will be generated (since it will no longer be burned), making this one of the best returns on investment. Energy sales are expected to account for 10 to 15% of mill revenues and between 25 to 35% of their operational cash flow.

2.6.8 Yeast

About 10% of the yeasts used in ethanol production, specifically in the fermentation of sugarcane, is recovered and dried to be used in the composition of animal feed. In 2008, yeast exports reached 32 thousand tonnes, generating revenues of US$ 16.80 million. This year was unusual because 15 thousand tonnes more could have been exported if contamination had not occurred. This problem has now been solved. In the domestic market, sales were US$ 11.09 million with 24 tonnes of dry yeast.

In addition to the yeast, additives based on sugarcane yeast (such as the cell wall) are marketed. In 2008, 13,400 tonnes of this product were exported, generating revenues of US$ 25.40 million. In the domestic market, 5,000 tonnes of additives were sold, representing a turnover of US$ 10.33 million. Therefore, sales of yeast in addition to its additives reached about US$ 21.41 million in the domestic market and around US$ 42.20 million in exports, totalling US$ 63.61 million.

2.6.9 Carbon credits

In terms of trading volume, Brazil ranks third among the country vendors, but it still has only 3% of the market. China and India are in first place with 84% and 4%, respectively. The amount traded worldwide in 2008 was 389 million tCO_2e, valued at US$ 6,519 million, and 14% less than in 2007.

The Brazilian participation in the carbon credit market occurs through the Clean Development Mechanism (CDM), because it is the only mechanism of the Kyoto Protocol that allows voluntary participation of the developing countries. The 68 Brazilian projects registered by the United Nations Framework Convention on Climate Change (UNFCCC) on the carbon credit market generated an estimated reduction of 3.45 million tCO_2e and a turnover of approximately US$ 25.35 million in 2008, using the average price in 2008 recorded by the voluntary market of $ 7.34 per tCO_2e. Of the 68 projects, 24 were from the sugar-energy sector, which generated an estimated decrease of 473.94 thousand tCO_2e, valued at US$ 3.48 million in 2008.

2.6.10 Bioplastics

Bioplastic is one of the innovations for the exploitation of sugarcane bagasse. If the planned investment really occurs, in a short time this product will be a very important item in sugar mill and ethanol plant portfolios. It is estimated that the demand for this new product has already reached 600,000 tonnes annually worldwide, although at a 15-30% higher price than the conventional product. According to the European Bioplastics Association, almost 331,000 tonnes of bioplastics are produced today, which is less than 1% of synthesised plastics produced annually. Brazilian bioplastics production is still performed on a minimum scale, inadequate for bringing product to the market.

PHB Industrial, a company controlled by one of the most important groups of sugar mills in Brazil, has in its industrial park one of the first pilot projects of the country. On a laboratory scale, the company can produce about 60 tonnes per year, which are currently exported to Japan, the United States, and Europe at an average price of US$ 5 per kg. However, very little of this material was sold in fact, and the majority was exported for developing applications with international companies. PHB Industrial is designing a plant to start operating on a commercial scale within 2 to 3 years. Media reports say the plant will eventually produce 10 thousand tonnes/year and will begin operations in 2010.

Braskem, a Brazilian petrochemical company, currently has a production capacity of about 12 tonnes/year in a pilot plant and has announced investments to start production in 2011 of approximately 200,000 tonnes per year. Dow Chemical reported the creation of the first ethanol hub that is slated to produce 350,000 tonnes/year starting in 2011. Copersucar, in partnership with the Belgian group Solvay, should produce 120,000 tonnes in 2010 (ABDI, 2008).

If investments for 2010 materialise, press reports estimate that the alcohol chemistry industry will require 650 million litres of ethanol annually. A large potential market signals unparalleled opportunities for the sector.

2.7 Facilitating agents

2.7.1 Brazilian development bank (BNDES)

The bank provided a sum of US$ 3,530.79 million for companies operating in the sugar-energy sector, thereby stimulating the development and maintenance of the industry.

2.7.2 Outsourcing of cutting, loading, and transportation of sugarcane (CLT)

Due to further industry consolidation, new groups have been taking the sugarcane business on a professional management basis with a focus on efficient operations and better financial allocation. This has created a demand for outsourcing services in the areas of cutting, loading, and transportation of the sugarcane from farms to the plants, favouring the entry of specialised companies in sugarcane logistics operations. In 2008, the outsourced CLT represented a turnover of US$ 916.32 million.

2.7.3 Sugar and ethanol road freight export

The resources dedicated to road freight for sugar and ethanol exportation totalled US$ 539.03 million. Of this total, spending on road freight for sugar exports in the central-south region was US$ 383.60 million, and the ports of Santos, in São Paulo State, and Paranagua, in Parana State, were the main export routes in 2008. Of that amount, freight exports of ethanol, including the ports of Santos and Paranagua, totalled US$ 155.42 million. The sugar export freight over the Brazilian road system costs approximately US$ 34.16/t, and ethanol freight costs US$ 34.76/m^3.

2.7.4 Tolls (road fees) for sugar and ethanol exportation

Revenues from tolls to export ethanol and sugar added an amount of US$ 79.96 million in 2008.

2.7.5 Cost at port (Port of Santos)

The revenue from the Port of Santos on customs clearances services, lifting, and supervision of loading the sugar and ethanol was estimated at US$ 213.52 million in 2008. Almost 70% of the entire Brazilian ethanol and sugar exports passed through the Port of Santos.

2.7.6 Research & development (R&D)

In 2008, US$ 79.15 million of resources were allocated to research on sugarcane, sugar, and ethanol production among public and private organisations.

2.7.7 Events

Currently, there are five important events in the sugar-energy sector that together mobilised US$ 5.32 million in 2008.

2.7.8 Specialty magazines

The major Brazilian specialty magazines in the sugar-energy industry earned US$ 3.99 million, with about 61 thousand copies printed.

2.7.9 Health care and meals

According to the Union of Workers in the Sugar and Food Industry, the São Paulo State workers receive health care and food benefits, totally or partially paid by the mills. The average monthly cost paid into health plans is US$ 33.00 per person. It follows that the health care segment in 2008 had revenues of about US$ 125.51 million for the sugar-energy sector. With regard to food, it is estimated that the São Paulo sugar mills and ethanol plants have spent about US$ 188.26 million, an average monthly cost of US$ 49.00 per person.

2.7.10 Wage bill/jobs

According to the Brazilian Ministry of Labour, the industry in 2008 accounted for 1.28 million formal jobs, with 481,662 allocated in the field of sugarcane cultivation; 561,292 in sugar mills for raw sugar production; 13,791 in sugar refining and milling; and 226,513 in ethanol production. This represents 2.15% of all Brazilian jobs, highlighting the importance of the sugar-energy sector. About 54% of all employees provided with jobs in the sector finished the year with no employment, which is typically repeated due to seasonal jobs that are generated at the harvest peak. The number of active jobs during the year was 588,826.

The figure of 1.28 million workers rises if we also consider the informal employment. The data show that the formal job numbers in the sugarcane industry have been growing, reaching 80.9% (Brazil), 66.5% (north-east), and 90.3% (central-south), and up to 95.05% in Sao Paulo State (IBGE, 2007). If informal employment is added to this, there are 1.43 million jobs in the sector. Considering also that every direct job generates two indirect, a figure of 4.29 million people placed in jobs related to sugarcane is reached.

In Brazil, 55% of the workers on sugarcane plantations are illiterate or of low education. The worst share is in the north-east region – with more than 80% of workers grouped in that category. In the central-south area the rate did not surpass 5%. In the sugar mills and ethanol plants, the proportion of illiterate and low-educated workers is slightly lower than on the plantations, but is still very high, highlighting illiteracy in the north-east, which in 2008 accounted for almost 20% of the workers. However, increased mechanisation has created a growing demand for more qualified professionals. A harvester replaces 100 workers with low skills, but it requires 10 workers trained in automation and mechanisation. Brazilian institutions are assisting in the training of this new profile of skilled worker required by the industry today.

The average income of the workers in the central-south region was US$ 578.00 per month, and in the north-east region it was US$ 362.00 per month, generating a national average of US$ 512.00. The national wage bill was US$ 738.33 million in 2008.

2.7.11 Taxes

Taxes were calculated using a 'weighted average rate', estimating the rate of goods in major states, considering the tax incentives and output. Only the tax on sale revenues were considered in this survey. In the case of the ICMS, we did not use the weighted average rate, but the 'interstate rate' of the central-south region. In the case of PIS/COFINS, we used the rates of 1.65% and 7.60%, respectively, with the exception of ethanol, which is taxed by a fixed value per litre. Moreover, in the case of the IPI we prioritised the rates of the most relevant products for each link. In order to estimate the tax for the sugar-energy sector, we classified companies as those organisations subject to the system of taxation on taxable income.

The total tax was calculated by summing the taxes generated in each link of the sugar-energy sector, from the sale of agricultural and industrial inputs to the sale of final products. To eliminate double counting and consider just the aggregate tax, we subtracted from this total the taxes generated in the first links (agricultural and industrial inputs).

The result of this estimate showed that the tax revenues in 2008 totalled about US$ 9,868.24 million, and US$ 3,012.84 million were generated by the sale of agricultural inputs and products. Thus, the aggregate tax in the sugar-energy sector was estimated at US$ 6,855.41 million.

2.8 Managerial implications and discussion

This chapter sought to examine the entire design of the sugar-energy sector, with an expectation of the dimension of that sector for the Brazilian economy. It is clear that the industry figures are impressive, with a turnover of over US$ 80 billion per year.

This is probably the most up-to-date picture of the production chain in Brazil. The material serves as valuable input for public and private decision-making, showing the interconnecting links between system components and the industry's enormous capacity to generate resources, taxes, and jobs.

The sugar-energy production chain has shown the world its potential to supply products on a sustainable basis, thereby ensuring that Brazil has one of the cleanest energy matrices – it is estimated that in 2015 ethanol will represent 80% of the total fuel consumed in Brazil by light vehicles. In addition, Brazil has a virtual monopoly on world exports of sugar, with nearly 50% of the global market, with the expectation of reaching more than 60% in 5 years.

This chapter also demonstrates that the sugar-energy sector accounts for a tremendous amount of resources, jobs, and taxes, and that its ability to internalise Brazilian development is very large. It is a sector of fundamental importance for the Brazilian economy.

3. Strategic plan for the sugar-energy value chain in Brazil

3.1 Introduction

To create a global market for food and bioenergy, it is necessary to develop expertise in the planning of agribusiness systems that will be increasingly transnational and that must be sustainable. According to Zylbersztajn and Neves (2000) and Batalha (2001), the agribusiness systems (chains) hold the following basic elements for their descriptive analysis: the agents, the relations between them, the sectors (inputs, agriculture, industry and distribution), the facilitating organisations, and the institutional environment. It is essentially a macro-analysis of the product flow from suppliers to final consumers.

In this process, to tackle the changes in the international business environment and growth opportunities for food and bioenergy, strategic planning is necessary. First of all, a closer look should be taken at the changes that are occurring regardless of the efforts of companies, governments and society. These changes in the macro-environment are grouped in political-legal, economic-natural, socio-cultural and technological terms and are applied here to the Sugarcane Agribusiness System (AGS).

3.1.1 Political-legal factors

Political-legal factors include ratification of the Kyoto Protocol and its effect on the pattern of energy consumption in the world's population; restrictions on the use of land (environmental impact) and water (for water recovery); requirements for waste, sewage and water management; targets for the reduction of emissions and incremental adoption of biofuels imposed by countries, through mandatory blending targets, on their organisations and society in general.

3.1.2 Economic-natural factors

Among the economical-natural factors are instability of the oil price; strong competition among issuers of renewable energy; growth in sales of *flex-fuel and hybrid cars*; addition of biodiesel to reduce sulphur emissions and improve the lubricity of the engine; addition of ethanol to petrol replacing MTBE (methyl tertiary butyl ether); opening of new markets for ethanol fuel, mainly the Asian market, new markets (electricity), and competition for biomass; chains of sustainable production (conditions and characteristics of agricultural production, respect for human resources involved, profits and distribution of results).

3.1.3 Socio-cultural factors

Social-cultural factors include growth in the segment of green consumers; biofuels: affirmation of the image of clean fuel; requirement for corporate social responsibility (social projects); growing concern for human health (level of pollution in big cities); search for better quality of life (better public transportation); defence of the national product: locally produced ethanol and biodiesel; quest for convenience and variety of products, where the consumer has the decision-making power (flex-fuel cars); enhancement of *fair trade* in purchasing decisions; growth in the consumption of organic products and requirement for traceability (GMOs).

3.1.4 Technological factors

The technological factors encompass improvements in the efficiency of flex-fuel cars and trucks (ethanol, petrol, natural gas, diesel and biodiesel); hybrid cars; hydrogen cells: fuel of the future; patenting of technology for the production of ethanol; technology of burning biomass and/or use of methane gas; major investments in the search for cellulose ethanol; integration of the ethanol plant and biodiesel (biorefinery), diversification of sources and energy production.

3.1.5 Strategic planning process

The construction and elaboration of a process for strategic planning and management of production chains should be the objective of each country. This process should prioritise the areas of coordination and institutional environment (laws), production and products, communications, distribution and logistics, and human resources, in order to define projects, strategic thinking, and change (Neves, 2007b).

The objectives of this section are to make an academic effort to incorporate the uncertainty, instability, and unpredictability of changes into the planning process of agribusiness systems; to consider various future possibilities in the formulation of objectives, guidelines, and strategies; and to ensure the sustainable growth of national economies.

We defined the following specific objectives:
1. Performing a macro-environment analysis (STEP analysis) for the sugarcane agribusiness system in Brazil.
2. Analysing the internal environment of the sugarcane agribusiness system in Brazil in five strategic areas, including its strengths and weaknesses.
3. Proposing a strategic agenda for this sector in order to collaborate in the discussions related to these issues.

3.2 An application of the CHAINPLAN method

It is estimated that by 2020 the world's food supply will have to be increased by 50% and that available areas for agriculture will become restricted, as will sources of water. An efficient logistical system is still a challenge for many countries around the world. It is difficult to predict how much biofuel will be needed, as it depends on car fleets and their evolution, industrial and human demand, institutional environments (% set by governments for the biofuel addition in oil), and consumer behaviour. Therefore, to deal with all of the environmental changes in the national and international business arenas, and the growth opportunities in the food, fibre and bioenergy productive chains, strategic planning should be focused on understanding the production chains.

Increasingly in Brazil it will be necessary to establish a strategic planning and management process for the nation's various agribusiness systems. For this, Neves (2007b) developed the CHAINPLAN method for 'Strategic Planning and Management of Agribusiness Systems', which has been applied to agribusiness systems in Brazil, Argentina, Uruguay and South Africa, among other countries. It consists of five stages: (1) the initiative of systems' leaders, (2) the mapping and quantification of the agribusiness system, (3) the formation of a vertical organisation, (4) the plan with strategic projects, and (5) the implementation of the plan. The CHAINPLAN method is summarised in Figure 20.

Figure 21 presents the details of step 4 of the CHAINPLAN method. This step is divided into 12 stages involved in the design of an integrated strategic plan for the agribusiness system for the next 5 to 10 years. Each stage is elaborated and detailed in Table 15.

Below, we delineate a strategic plan for the sugarcane agribusiness system, in order to cooperate with existing organisations in the system, that have been working in this direction. It is the result of proposals that have been made by the authors over the last 10 years as well as suggestions for the future.

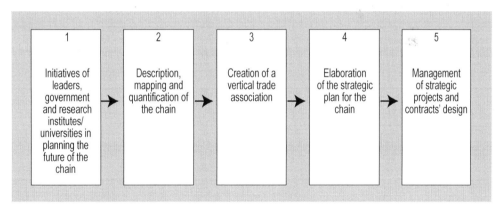

Figure 20. The CHAINPLAN method for implementation of strategic planning and management in production chains (Neves, 2007b).

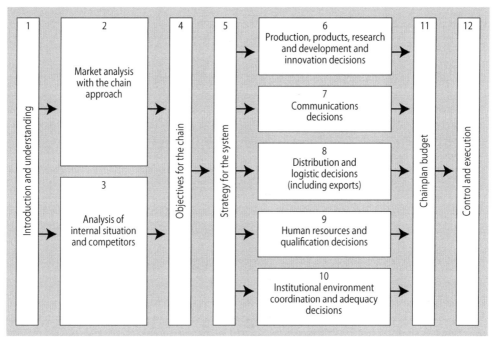

Figure 21. Summary of step 4 of the CHAINPLAN plan (Neves, 2007b).

Table 15. Guidelines for demand-driven strategic planning and management of the production chain (Neves, 2007b).

Stage	What has to be done
Phase 1 – Introductory	
1. Introduction and understanding	Verify whether the chain has made other plans and study them
	Verify the planning method of the studied chain
	Verify which teams will take part in the process
	Study plans made for production chains in other countries, for benchmarking
	Identify a member of the team who could promote relations with other chains
	Finally, in the case of chains with sophisticated planning processes, it must be verified how this model can help, and be gradually adapted to, the existing model

Table 15. Continued.

Stage	What has to be done
Phase 1 – Continued	
2. International market and consumer analysis with chain approach	Address threats and identify opportunities from the so-called uncontrollable variables (possible changes in the legal-political, economical-natural, socio-cultural, and technological environments) in the domestic as well as the international market
	Understand existing barriers (tariff and non-tariff) and check collective actions to reduce them
	Analyse the final and intermediate (dealers') consumer behaviour and their purchasing decision processes
	Analyse opportunities to fit environment, fair trade, sustainability and sustainable development goals
	Analyse opportunities to fit the national and international labour institutional environments
	Set up an information system to support informed decision-making
	Identify the main national and international competitors
3. Internal situation analysis and global competitor benchmarking	Identify all the strengths and weaknesses of the chain
	Map contracts and existing forms of coordination
	Describe the existing structures of management and their transaction characteristics
	Make an analysis of main competitors
	Analyse the value creation, resources, and abilities of the chain
	Analyse the critical success factors of the chain
	Select, among the chains (which may or may not be a competitor), where the benchmarks (sources of good ideas) will be
4. Objectives for the chain	Define and quantify the major chain objectives in terms of production, exports, imports and sales to achieve sustainable growth and to develop solutions for the weaknesses
5. Strategies to reach proposed objectives	List the major strategies (actions) that will be used to reach the considered objectives in item 4 in terms of positioning, exports, value capture and market segmentation

Table 15. Continued.

Stage	What has to be done
Phase 2 – Plans of strategic vectors: production, communication, distribution channels, qualification, and coordination (institutional adequacy)	
6. Production, products, R&D and innovation projects	Analyse productive potentials and production capacities
	Map and plan for production risks (sanitary and others)
	Analyse products and product lines, as well as complementary product lines for expansion decisions
	Develop innovation opportunities in the chain, and in the launch of new products
	Identify opportunities to set up national and international innovation networks
	Foster partnerships with universities and with medical and health institutions
	Detail all offers and potential services
	Make decisions related to the joint construction of brands and labels for the system use
	Analyse and implement the certification process for the chain
	Ensure product adequacy with respect to the rules and institutional environment
	Ensure environmental sustainability
	Make packaging-related decisions (labels, materials, design)
	Calculate recurrent investments at this stage
7. Communication projects	Identify the target public for the communication (messages from the production chain)
	Develop goals for this communication (product knowledge, product reminders, persuasion, among others) and try to define the unique positioning and message that will be generated by the chain
	Define the communication tools to be used; that is, define advertising or public relations strategies to boost sales, among other things
	Produce films and international materials that benchmark those already used in other production chains
	Review communication actions and determine the annual promotion budget, including all network agents
	Indicate how communications results will be measured so that the chain continues to learn more about the best tools for achieving return on investments

Table 15. Continued.

Stage	What has to be done
Phase 2 – Continued	
8. Logistic and distribution projects (including exports)	Analyse the product distribution channels and search for new ones
	Analyse the possibilities of value capture in the distribution channels
	Identify possible needs of international dealers and consumers so as to be able to adapt the existing services
	Define new ways to penetrate the markets (through franchising, joint ventures, and other contractual forms, or through vertical integration)
	Determine annual budget for distribution
	Verify how distribution actions can be carried out together with other chains
9. Enabling decisions in the productive chain/human resources	Conduct training in management for the chain participants
	Conduct training in cost control and use of technologies
	Conduct training in national and international sales
	Transmit information from technological centres/research
	Conduct training in food production
	Offer technical assistance to improve properties
10. Institutional environment coordination and adequacy projects	Develop projects to finance the chain
	Develop basic infrastructure-improvement projects
	Develop projects to increase consumption in government programmes
	Develop programmes for isolated productive areas
	Push for tax reduction in the production chain project
	Strengthen export activity through export promotion agencies
	Support laws that provide incentives for the use of technologies (fiscal incentive, etc.)
	Develop a product and product name standardisation project
	Promote more transparency in legislation referring to products and process projects
	Develop proposals for chain conflict solutions
	Ensure coordination in the development of contracts and proposals
11. Strategic projects consolidation	All projects generated in steps 6 to 10 will be consolidated and priorities will be set
12. CHAINPLAN budget	Budget of every project, which contains costs and total budget

3.3 Understanding

In this section, a historical and broad understanding of the industry is outlined. It is believed that the sugarcane system is widely known in Brazil. There are several technical and market publications, and this paper aims to add to this existing knowledge. Furthermore, the sugarcane industry is characterised by strong and representative industry associations, including the industrial, production, and distribution links, along with trade unions and others who have done significant development work. A strategic plan should include organisations, companies, and other public and private agents gathered especially for this purpose.

3.4 External analysis: opportunities and threats

There are many steps that must be taken in the external analysis, as can be seen in Table 15. In this section, we seek to contribute to the macro-environmental analysis using the 'PEST or STEP analysis' tool, which is commonly employed in the strategic planning literature. It considers the main uncontrollable factors in a production system that bring opportunities and threats. These are the environmental factors, already mentioned in Section 3.1: the political-legal, economic-natural, socio-cultural and technological factors (Neves, 2007a; Campomar and Ikeda, 2006; Jain, 2000; Johnson and Scholes, 1997).

To develop multiple scenarios of a theme that involves the energy future of the world, and dealing with an industry known for its high-risk and fairly long-term investment projects, various macro-environmental changes are analysed. The main 'uncontrollable' factors to be considered in the analysis of the energy sector in the world today are presented in Table 16, which separates these changes into environmental opportunities and threats.

3.4.1 Biofuel drivers for investments in Brazil

'Biofuels offer tremendous opportunities for farmers, rural communities and developing countries to improve local development' (Dallas Tonsanger, Sub-Secretary of Agriculture and Rural Development, USA)[3]

Oil prices: *current and future oil prices and availability of oil reserves*

The use of biofuels is stimulated not only by environmental issues, but also by oil prices. The problem of oil price fluctuations makes it increasingly hard to base the economy on this kind of energy source. Between 1998 and 2007, the price of the oil barrel increased more than

---

[3] Speech made at the Ethanol Summit, Sao Paulo, Brazil, 02/05/09.

500% (NYMEX, 2007). On February 19th 2008 the barrel reached US$ 100.00 for the first time in history. Nowadays the price of an oil barrel stands between US$ 50.00 and US$ 80.00.

Pressure on prices comes mainly from the perspective of complete reserves depletion. Some studies indicate that the reserves will run out in approximately 40 years (BP, 2006). Despite the discovery of new reserves, they will be unable to meet the long-term growth in energy demand. According to IEA (2006), based on the current trends of global energy, demand will rise by up to 53% by 2030. Over 70% of this increase comes from developing countries, led by China and India. Imports of oil and gas in the OECD and developing Asian countries are growing even faster than demand. Global oil demand is set to reach 116 million barrels a day (b/d) in 2030, compared to 84 mb/d in 2005.

Another risk factor, in addition to the unstable prices and the possibility of scarcity, is the fact that the largest oil reserves are situated in unstable regions. The main suppliers of oil remain in the Middle East, with 62% of the world's reserves, followed by Europe and other regions of the Asian continent (BP, 2006).

From this perspective, will biofuels be viable? According to UNICA (2007) projections, with oil prices above US$ 80.00 per barrel, biodiesel becomes viable. For ethanol, the scenario is even better: oil prices just over US$ 40.00 a barrel make Brazilian ethanol derived from sugar-cane viable.

Transport dependency on oil: *energy demand in the transport sector, transport consumption of fossil fuels comparative to other sources and the participation of the transport sector in the world's energy matrix*

As far as worldwide fuel consumption is concerned, according to WBCSD (2004) the transport sector is expected to increase its share in oil products (from 56% to 62%) due to the increase in consumption of 60% within the period 2000-2030 (2.1% a year). Therefore, fossil fuels should remain a key energy source for transport, despite the advances in renewable and less carbon-intense fuels (LPG, ethanol, biodiesel and hydrogen). Changing this scenario will require investments in R&D (Research and Development) as well as in the image of biofuels as a clean, safe and low-cost energy source.

In North America petrol represents more than 50% of the total energy demand for transport while diesel represents a little more than 20%. West Europe shows a different consumption pattern as both diesel and petrol each respond to about 37.5% of the sector demand. In Asia, petrol is used more (45%). Thus, the American and Asian markets are the most promising for Brazilian ethanol (WBCSD, 2002).

As regards the share of the road transport categories in fuel consumption, light vehicles and trucks represented over 60% of the demand in 2002. However, light personal vehicles

Chapter 3

Table 16. Summary of opportunities and threats in sugarcane AGS (agribusiness system).

Political-legal	Economical-natural
Opportunities	
Addition of ethanol to truck engines	Growth in population and increase in wealth (China
New emission-reduction targets and growth of the	and India), increasing consumption
carbon-credit markets	Growth in the consumption of sugar (products/food
General tax incentives for biofuel production	that use sugar)
Development and internalisation of biofuels market	High oil prices
in developing countries, with the advancement of	Growth in flex-fuel vehicle fleets
new projects (biofuels and feedstock production)	Export of technologies and biofuel facilities from
in degraded areas	actual producers' countries to new ones
Addition of ethanol in different countries,	New and high flows of foreign direct investments in
replacement of MTBE used in petrol to meet	biofuel industries
environmental agenda	Loss of production in some countries, generating
Addition of biodiesel in different countries, the	opportunities for others
lower level of sulphur emissions and greater	Emergence of new producers (Caribbean and Asia)
lubricity to the engine	Focus on core competence (biofuel industry),
Alliance of developing countries with developed	independent supply of feedstock with better
countries to obtain preference for imports and to	income distribution
avoid competition with food production	Good agricultural practices like crop rotation
Prohibition on burning sugarcane, generating more	(food and energy), causing an increase in food
energy to crush and ethanol facilities	production in the areas of renewable energy crops
Brazilian ethanol as advanced biofuel in the USA	Land availability for expansion of the biofuel sector
New institutional framework for electricity	in developing countries
Environmental zoning in Brazil	Positive energy and carbon balances for all biofuel
New institutional framework for distribution of fuels	sources
	Value of positive externalities

Food and fuel

Social-cultural	Technological
More awareness of global warming	New technologies enhancing flex-fuel vehicle efficiency
Migration of people to cities (e.g. China), requiring processed food and high volumes of fuels	New machines for harvesting crops
Image of renewable and clean fuel	Generation or expansion of cellulosic ethanol use (biobutanol, hydrolysis)
Defence of sustainable biofuel productive chains	Genetic modification of energy crops for resistance to dry weather and diseases
Acceptance of GMOs	Use of satellites and precision agriculture (GPS)
Social movements in areas of organics, fair trade, nutraceuticals and cosmetics	Research with fertilisers (varieties that use less fertiliser)
Inclusion of small farmers	Use of biofertilisers from by-products
Generation of green jobs and income	Integration of biodiesel and ethanol facilities
	Focus on energy efficiency (hybrid cars, reducing the weight of cars), allowing the use of renewable energy (ethanol, biodiesel, biomass)
	Privatisation/public-private partnerships in infrastructure facilities

Table 16. Continued.

Political-legal	Economical-natural
Threats	
Social-environmental barriers to biofuel imports	Strong growth in the hybrid vehicle fleets
Lack of international law on biofuel standardisation for export (in the world market)	Lack of machines and equipment for expansion of industrial capacities
Stricter labour and environmental laws for biofuel production	High agricultural commodity (feedstock) price fluctuations
The oil companies, the local producers, and the ethanol lobbies against imported ethanol	More powerful diseases or pests
Slow and tendentious legal environment (contractual hold-up problems, delays in justice, bureaucracy, etc.)	Climate change bringing reduction in the available lands
Lack of regulatory stocks of biofuels in countries (to avoid fluctuation in commodity prices)	Lack of agro inputs (fertilisers mainly)
End of the tax incentive programmes in the long term (breaks)	Concentration of the biofuel sales in a few major markets (USA, EU) or companies (e.g. BP, Exxon, Chevron, Shell, Petrobras)
Tax inequality through value chain and states of Brazil	Inflationary process in food prices
Conflict 'pre-salt' investments vs. bioenergy economy in Brazil	Competition between biofuel industries and alternative distribution channels by ownership of by-products (agricultural residues)
Petrol price control in Brazil and Petrobras monopoly	Lack of credit/funding lines with easy access
	Small environmental services markets in Brazil

only represented 50%. Due to their high per capita income, the developed countries have the largest fleet of light duty vehicles (WBCSD, 2004). Improvements in per capita income usually mean more vehicles.

Governmental biofuel incentives: *subsidies and tax incentives*

Much of the world's production of biofuel needs some kind of incentive such as subsidies or tax exemption to make the prices economically viable in comparison to fossil sources. In this sense, OECD data (2005) (average 2002-2004) presents the countries that support their producers the most (in terms of % of the growers' gross revenues which comes from governmental support). These are Japan (58%), the European Union (34%), Canada (22%), Mexico (21%) and the USA (17%). In Brazil only 3% of the producers' revenues come from federal support, due to the subsidised interest rates originating from agricultural debt renegotiation.

Social-cultural	Technological
High supply and use of public transportation	Sweeteners and other bioenergy sources
High migration flows of people to developed countries	New technologies generating more competitive energy (hydrogen)
Image of jobs generated by the energy crops employed in the harvest in developing countries (sugarcane, palm)	Growth in the fleet of natural gas or hybrid vehicles
	Deficient infrastructure for distribution of agricultural production from new frontiers (internal logistics)
Image of land occupation generating competition with food	Low investments in R&D in developing countries
Image of the 'monoculture'	
Growth of NGOs with destructive purposes (bioterrorism)	
Strict requirements of social-environmental certification	
High cost of certification	
Mechanisation vs. unemployment in agriculture	
Number of different seals	

Tax reduction can come in many guises. It can be applied to the production and trade of biofuels, flex-fuel vehicles, as well as in engine conversion services that allow the use of ethanol, biodiesel, or blended fuels. Some governments also used to offer special financial opportunities for projects that involved biofuels.

The USA has a combination of federal, state, and local subsidies that cover each transaction of the whole production chain (industry, storage facilities, distribution centres and ethanol end-users) and also the purchase of clean vehicles. In 2003, the European Commission authorised Member States to give tax exemptions for ethanol and biodiesel producers (Steenblik, 2007).

The Brazilian government offers tax deductions for biodiesel companies that buy a minimum percentage (50%) of their feedstock from small growers, who produce some specific oilseeds (*Jatropha curca*, castor oil and palm) in the northern and north-eastern regions of Brazil (Probiodiesel, 2007).

Some countries also reduce the export tariffs for biofuels in an attempt to stimulate internal production. Argentina, for instance, has different tariffs for the products of the soybean value chain. While soybean meal and oil exports are taxed on 24.5% of total revenue, the exports of biodiesel are taxed on 5% (Mathews and Goldsztein, 2007). With this policy, the government can stimulate production without necessarily having to depend on internal demand.

Governmental biofuel restrictions: *barriers* (ad valorem *and specific import taxes, import quotas, fuel standardisation, and certifications*)

While governmental incentives seek to encourage domestic production of biofuels, there are some restrictions which protect local growers from foreign competition. These restrictions may be in terms of fuel standardisation, requirements for some specific productive skills, social and environmental certifications, import quotas, and import tariffs, among other things.

Import tariffs are the most relevant alternatives for restriction, such as *ad valorem* tax and specific tariffs. As IEA (2004b) data shows, Australia, a major producer of sugarcane ethanol, has a specific tariff of US$ 0.24 per litre of ethanol. The European Union charges import duty of US$ 0.10 per litre (prospect of environmental certification requirement), while Canadian importers pay US$ 0.07 per litre (strong ethanol producer from maize), the same value per litre as in Brazil. In the USA, the world's biggest ethanol market, the import tariff is US$ 0.54 per gallon – the mandate expires in 2010 but is likely to be extended.

Tariffs are not limited to final products. Some countries impose import tariffs on raw materials used in the production of biofuels.

Clean vehicle adoption: *the size of vehicle fleet and its growth rate, adoption rate of hybrid cars vs. flex-fuel cars by the main car buying countries, possibility of introducing cars that run on hydrogen fuel cells, growth in the number of LDVs (Light Duty Vehicle), and number of cars per inhabitant*

The developed countries hold the largest share of the world's fleet. However, the situation in developing countries requires more attention. A Goldman Sachs' forecast indicates that by 2040 China and India will have, respectively, 29 and 21 cars for every hundred inhabitants, totalling more than 700 million cars.

The automobile sector makes remarkable investments in R&D in the area of using alternative sources of energy in engines. Two cases serve as a reminder: the hybrid car (a car that combines a petrol engine with an electric battery) and the flex-fuel car (an engine that can be fuelled with petrol, ethanol, or a blend of both).

The world's largest fleet is in the USA. There are around 250 million vehicles running on American roads and another 15.5 million are expected to join them in 2008 (RFA, 2008). The production of E85 (85% ethanol and 15% petrol) cars is growing faster than those of other vehicles. In 2005 alone, E85 flex production grew 16% against a 5% growth in the production of vehicles that burn exclusively fossil fuels (OICA, 2007). According to the Renewable Fuels Association (RFA, 2008), the USA already has 7 million E85 flex vehicles in their fleet. The biggest obstacle to the spread of this technology throughout the country is the paucity of fuel stations that offer the product. Less than 2% of the 170,000 American fuel stations have E85 pumps.

Flex-fuel cars have been adopted in Brazil since their launch in 2003. This new technology drastically changed consumer attitudes since it minimises some of the risks associated with the exclusively ethanol car, like shortage of fuel and high prices in mid-season. The effect of FFVs (flex-fuel vehicles) on car sales was intense and fast. In 2003 FFV sales represented less than 7% of all cars sold in Brazil. Nowadays FFVs account for over 90% of total sales and already represent 40% of Brazil's light vehicle fleet. By 2015 the Brazilian fleet is projected to have 30 million vehicles, of which 19 million should be FFVs (ANFAVEA, 2010 and UNICA, 2010).

This situation is the result of convenience and product availability for consumers. In Brazil all 35,000 fuel stations are supplied with ethanol, and the biofuel made from sugarcane has already substituted an enormous volume of petrol. In 1988, when the ProAlcool programme was at its peak, ethanol consumption represented 57% of the total light vehicle fleet demand for fuel. Currently, the ethanol share in the fuel market stands at 54.5% (ANP, 2010).

In the long term, plug-in hybrids, biofuels from cellulosic materials and hydrogen fuel cells are interesting alternatives, but they all require major advances in R&D to reduce production costs.

Generally, ethanol and biodiesel prices at the pump are influenced by the prices to producers, the volume added to petrol according to the mandatory blending target, the logistics and the distribution costs and taxes. However, the major influence on biofuel consumption is actually the price of other fuels (mainly petrol and diesel), the vehicle consumption levels and the characteristics of the fleet (release of flex or hybrid vehicles, prohibition of diesel engine light duty vehicles, etc.).

Feedstock production capacity: *productivity and production cost of biofuels around the world, rate of irrigation use, available land vs. occupied land, current and future price of feedstock (sugarcane, grains and vegetable oils) as a consequence of food consumption, food and fuel competition.*

In several countries the production of ethanol and biodiesel still depends substantially on subsidies for its survival in the market. Most of the time, high costs are associated with less than optimum production rates – in comparison with substitutes – and with low use of the by-products (agricultural residues).

There are considerable differences in ethanol productivity according to the raw material used as well as the production location. Comparatively, Brazil is the country with by far the best yields, producing an average 6,800 litres of sugarcane derived ethanol per hectare. In contrast, the EU produces 5,400 litres/hectare of sugar beet ethanol and only 2,400 litres/hectare of wheat ethanol, India 5,200 litres/hectare (also from sugarcane), the USA 3,100 litres/hectare (from maize) and Thailand also 3,100 litres/hectare (from tapioca). This fact helps Brazil produce the cheapest ethanol in the world for a price of US$ 0.22/litre. In the USA, the biofuel made from maize costs US$ 0.30/litre and the Europeans produce ethanol for US$ 0.45 (grains) and US$ 0.53 (sugar beet) (F.O. Licht's, 2007).

Investments in R&D to improve agricultural production (irrigation methods, genetic improvement in seeds, management skills, improvement in fertilisers and others) are strategic actions for consolidating biofuels as an alternative energy source. On the other hand, there is also a limit to the amount of available agricultural land for biofuels and a trade-off between food and biofuel production. Developed countries are at a disadvantage because most of their agricultural land has already been explored and such competition will be inevitable. FAO's (2007) data shows that only a few countries still have available land for agriculture conversion. Brazil has the most available land with 394 million hectares, of which only 66 million are being used. After that comes the USA with 81 million hectares of unused land, Russia (88 million), the EU (61 million), China (42 million), Australia (37 million), Canada (30 million) and Argentina (44 million). Despite its huge territory, all of India's 169 million hectares are already occupied.

Biofuel production capacity: *construction of new facilities, increase in ethanol productivity with hydrolysis (cellulosic ethanol) and usage rate of by-products*

In recent years, the construction of biofuel facilities has grown intensively. The reason for this, in many countries, is the increase in domestic market demand with blending targets and also the possibility for potential exports. A more in-depth analysis carried out by the Inter-American Development Bank (Rothkopf, 2007) points out that in 2005 investments in bioenergy (ethanol, biodiesel, biomass for electricity and some others) reached US$ 2.66

billion and only one year later, in 2006, this amount was 7.9 times bigger, reaching US$ 21 billion.

There is also the possibility of using the hydrolysis process to obtain ethanol. Hydrolysis allows the ethanol to be produced from any possible source of cellulose. In the case of maize and sugarcane, the hydrolysis process will use residues such as leaves, straw and bagasse (from sugarcane). Nowadays, some of these by-products are under-used or even discarded. This industrial process is, however, still in its early stages of development.

This new technology would increase ethanol production worldwide using the same agricultural lands. In 2005, the production of conventional ethanol in Brazil was 85 litres/ tonne of sugarcane or 6,000 litres/ha. In 2015, the conventional production will reach 100 litres/tonne or 8,200 litres/ha, and the production by hydrolysis 14l litres/tonne or 1,100 litres/ha. In 2025 conventional processes are expected to produce 109 litres/tonne or 10,400 litres/ha, and hydrolysis some more 3,500 litres/ha (Leal, 2006).

According to the National Renewable Energy Laboratory (NREL, 2006), cellulosic ethanol will be the solution for increasing yield and enabling production to meet the global demand for fuel. Some countries like Brazil have already begun using residues from the fields as a source of energy (bagasse and leaves) and biofertilisers (vinasse). This results in an increased yield and lower production costs even though collecting these residues comes at a cost.

Social improvement: *capacity for job creation, minimum viable size of farm by feedstock (familiar agriculture × entrepreneur agriculture), and mechanisation rate of harvest*

Some researchers suggest that biofuels could play an important part in helping poor countries to diversify business and ensure sustainable growth. According to Zarrilli (2007), several countries that have implemented biofuel development programmes have shown noticeable growth in job creation, most of them created in the rural areas but also in other links throughout the productive chain.

According to Poschen (2007), the senior International Labour Organization's specialist on sustainable development, the number of jobs created in the renewable energy sector will double by 2020, creating approximately 300,000 new jobs. In the early phase of the bio-ethanol programme in the USA, around 147,000 jobs were created in different sectors of the economy.

In 2008, the Brazilian sugarcane industry took on 1,283 thousand Brazilian workers, of which 481,662 were in the ethanol industry, 575,083 in the sugar industry and 481,600 in the sugarcane production. That represented a total increase of 99.6% in the number of jobs since 2000 (RAIS, 2008 cited in Moraes, 2009).

Producing different biofuels requires the use of different production methods and thus the creation of varying numbers of jobs. Biodiesel production presents a better scenario for job creation, since other crops (e.g. palm, *jatropha,* castor bean) can also be produced by small farmers. In Brazil, each 6 ha of palm creates one job (EMBRAPA CPAA, 2007). In the cases of maize and sugarcane, the development of small producers is not significant since this agricultural activity demands a high scale of production and mechanisation in order to be economically feasible. As for Brazil, a sugarcane producer must have a planted area of at least 500 ha to be able to mechanise harvesting without incurring economic losses (Mello and Paulillo, 2005 cited in Camargo, 2007).

At the same time, men are being replaced by machines for harvesting. Sugarcane and maize harvesting can be mechanised, while this is not yet possible for palm. According to UNICA (2010), a potential scenario where 100% of the sugarcane harvest in the State of Sao Paulo – the largest producer of sugarcane in Brazil – and 50% in the rest of the country is dealt with by machines, would cause a reduction of 165,000 jobs compared to the number of workers in 2000. On the other hand, there is a growth in the demand for more qualified workers in the sugarcane field, in the sugar and ethanol industry and also in other sectors such as machines and service suppliers. Currently, machines already harvest more than 50% of the sugarcane produced in the state of Sao Paulo.

It is also a fact that innovations in growing sugarcane and grains have created better working conditions around the world and have reduced the negative impact on the environment. As Balsadi (2007, cited in Camargo, 2007) states, the results of such innovations in Brazil can be seen not only in job regulation, but also in the elimination of child labour, the increase in literacy rates and the rise in salaries and benefits.

Environmental improvement: *energy balance, potential GHG (Green House Gas) emission reduction (carbon sequestration or avoided emissions), and reduction cost (US$/tonne CO_2e)*

One of the most important reasons for biofuel consumption is its environmental significance, especially considering the urgent necessity for reducing GHG emissions (mitigation) as a means of avoiding bigger climate changes and their potentially catastrophic consequences.

The transport sector bears the greatest responsibility for GHG emissions related to energy activity. A close look at both current and projected CO_2 emissions from transportation reveals that road transportation leads the emission ranking in the present and future (currently rank 3/4) (IEA, 2005 and WBCSD, 2004). In this case, adding biofuels to fossil fuels has a tremendously important role to play in diminishing the negative impacts of the transport sector on the environment.

A study by the World Watch Institute (WWI, 2006) shows that energy balance (renewable energy in the biofuel divided by fossil energy used to produce it) is positive in the case of biofuel production and use (the whole production chain). However, there are several differences between feedstock for ethanol: maize in the USA (1.4), sugarcane in Brazil (8.3), wheat and beet in Europe (2). The same analysis is also made for biodiesel: oil palm (9), residues of vegetable oils (5.5), soybean (3) and colza (2.5). For example, the sugarcane chain in Brazil and the oil palm chain in Indonesia and Malaysia do not use fossil energy (or only a minimum) in the industrial process, only residues, ensuring great sustainability in the process and reducing GHG emissions.

NIPE/UNICAMP professor Isaías Macedo has recently published a paper in *Biomass and Bioenergy* (Macedo *et al.*, 2008) in which he states that the superiority of Brazilian sugarcane ethanol, in terms of energy balance throughout the whole chain, will cause the level of 8:1 (renewable energy: fossil energy) to reach 10:1 by 2020 with the hydrolysis process of bagasse and leaves, and with the trade of electricity. As far as carbon balance goes (avoided emissions and produced emissions), in a scenario for 2020 the use of E100 FFVs would reduce 2,259 tCO_2e/m^3 and the use of E25 petrol vehicles would reduce 2,585 tCO_2e/m^3. This paper and its data are largely based on the systematic analysis of 44 Brazilian sugar and ethanol mills which process about 100 million tonnes of sugarcane every year.

A report by the International Energy Agency (IEA, 2004a) shows that biofuels can contribute significantly to reducing the amount of CO_2 emissions. Ethanol from sugarcane (Brazil) contributes about 85% of the reduction, ethanol from grains (USA and EU) contributes 30%, and ethanol from beet (EU) contributes 45%. Cellulose ethanol (IEA) with 105%, creates the highest level of CO_2 reduction. In comparison to diesel, biodiesel virtually halves the volume of CO_2 emitted. At the same time, in terms of the cost of CO_2 reduction (US$/ tonne CO_2) ethanol from sugarcane (Brazil) is the cheapest option of all the biofuels (less than US$ 40.00). After that, there is American ethanol made from maize (over US$ 45.00), ethanol from sugar beet in the EU (US$ 300.00 and ethanol from grains in the EU (more than US$ 600.00)

Hence, among the technological possibilities for reducing energy consumption and consequently GHG emissions, are the following suggestions: reducing the weight of vehicles (lighter materials, improved aerodynamics), improving engine efficiency (direct injection, hybrid vehicles), and a greater use of alternative fuels (biofuels, natural gas, hydrogen/fuel cell and batteries). There is no doubt that the adoption of biofuels is the best option for ensuring that the transport sector plays its role in reducing GHG emissions.

Nevertheless, some critics say that not all biofuels are as sustainable as they claim to be. In order to avoid confusion on sustainability of biofuels, the market can develop instruments such as certification on sustainability. The main initiatives on the certification of biofuels produced up to now have come from national governments, the private sector, non-governmental and international organisations. The process of certification starts with a definition of the sustainability principles that address the environmental, social and economical matters; establish effective criteria; create clear and precise indicators that allow the quantification of the benefits that are to be reached; define an economically viable methodology; and organise monitoring systems (Mathews, 2008).

The most important criteria and standards that have already been developed refer to the energy and GHG balance, the protection of biodiversity, the competition with food production, leakage[4], and social and environmental matters. As far as the operational structure goes, the majority of certification processes opt for tracing systems. These systems usually differ among themselves in the verification tools and in their relation to national policies which are mandatory for some and voluntary for others (Van Dam *et al.*, 2006).

Lately, the international market for biofuels has opened up, especially to anhydrous ethanol due to government policies on adding biofuel to petrol. Some countries already have approved mandatory blending targets, while others have only authorised blending.

Since the international demand for biofuels depends on the establishment of mandatory blending targets by countries, scenarios are likely to be designed for biofuel demand according to this institutional environment.

Table 17 summarises the biofuel demand scenarios post-2020, according to the authors. In order to come up with these scenarios, we had to define the key variables (key drivers) of analysis, which are the variables of the environment that have greater power to influence the focal issue (biofuel demand), despite its uncertainty, and the capacity to generate contrasting scenarios according to its different final states.

Among the big producers and consumers of biofuels, the strategic objectives are very clear. The USA's recent approval of the New Energy Bill, which demands consumption of 36 billion gallons (or 136.8 billion litres) by 2022 in order to replace 15% of the domestic petrol demand, highlights their concerns about energy security in times of unstable oil prices. The EU's intention to add 10% biofuel to the road transport sector by 2020 should prevent 35% GHG emissions for each unit of biofuel in comparison to petrol and diesel, and underlines their concerns about climate change.

These two strategies are alike in that they support domestic agriculture. While the USA aims at making maize ethanol viable, the EU is trying to ensure the production of biodiesel from colza. Nonetheless, the international biofuel market cannot rely on these two blocks alone.

It is estimated that by 2025, we will have to increase the supply of food in the world by 50% (Boularg, 2007), while there are few agricultural areas available (3.23 billion hectares). We

---

[4] The term *leakage* refers to the net balance of GHG emissions that occurs from project borders.

are also faced with the question of how to allocate areas to bioenergy. This is difficult to predict, because it depends on fleets of cars and their development, industrial demand, individual demand, institutional environments (% set by governments for adding biofuel), and consumer behaviour.

However, if a barrel of oil costs less than US$ 40, and if there is less pressure in relation to global warming, the 'tsunami' of biofuels may seriously hinder the emergence of new technologies that provide ethanol and biodiesel and the inflation in food demand. We do not believe this will happen; instead we will see the following impacts on agribusiness:
- greater land exploitation;
- internationalisation of agribusiness;
- transfer of income from society to farmers;
- improved image of agriculture;
- less resistance to GMOs;
- serious problems in the supply of fertilisers; and greater use of biofertilisers;
- problems with the provision of crop protection chemicals, machinery and industrial equipment;
- acceleration in the professionalisation of, and concentration in, agribusiness.

3.5 Internal analysis: strengths and weaknesses

In addition, the third part of the strategic plan requires several analyses that compare the world's main producers and exporters (Australia, India and Thailand, among others related to sugar; the USA related to ethanol), and provide an understanding the competitive benchmarks. This is followed by an internal analysis, which determines and highlights the strengths and weaknesses. The idea is that strengths can be enforced and weaknesses can be worked out through the implementation of strategic projects. The analysis areas or themes are divided into five steps, as in the CHAINPLAN method. First, there are the innovation, research, and production issues, followed by the issues related to production system communication with clients and consumers. The third dimension addresses issues such as distribution and logistics, and the fourth analysis involves training. The last step is related to coordination aspects of the productive system and the institutional environment. These points are outlined in Table 18.

Table 17. Planning TNC foreign direct investment in Brazil – biofuel demand scenarios post-2020.

	'Pessimistic' scenario: countries reduce current targets. USA (blending targets 15% → 10%), China (blending targets 15% → 10%) and EU (10% → 5.75%)
1. Oil prices	Discovery of new wells Increase in production Barrel at US$ 40
2. Transport oil dependency	Consecutive economic crises Less credit Investment in clean public transport and fewer personal vehicles Investment in rail, water and air transport
3. Governmental biofuel incentives	Countries with blending targets but no subsidies or tax incentives Rule on domestic regulation only (no international standardisation) Priority for food production
4. Governmental biofuel restrictions	Rise in protectionism Strong international reaction to 1st generation biofuels from developing countries
5. Clean vehicle adoption	Predominance of non-combustion vehicles sales (hydrogen + electric) Less than 50% of flex-fuel or hybrid vehicles in the fleet

'Expected' scenario: countries maintain current targets	'Optimistic' scenario: rise in current targets + adoption by other countries such as Russia and Japan
Steady production (at recent levels) Low investments in prospecting new wells Barrel at US$ 80	Fall in production by main suppliers, located in unstable regions (production below historical levels) Low investments in discovering new wells Barrel at US$ 120
Maintenance of economic prosperity, but with lower growth rates than recent years	Rise in economic prosperity Maintenance of current economic growth rate and personal and commercial vehicle sales
Maintenance of current tax incentives and subsidies Move towards international standardisation Certification and regulation in order to transform ethanol and biodiesel into commodities	Rise in subsidies and tax exemptions Considerable rise in efforts to promote standardisation Social and environmental certification and regulation
Maintenance of agricultural protectionism in favour of local producers Growth of preferential markets (USA with CBI – Caribbean Basin Initiative, EU with EBA Agreement (British Sugar/Illovo, investments in Africa) and the SD&G Agreement (14 countries, mainly in Latin America) The USA maintains import tariff on ethanol Some EU countries break the rules and the non-tariff barriers imposed by the Commission, in order to achieve their own objectives	Production concentration in more competitive countries (mainly in the South) Northern countries prioritising food production Strong growth of free market
Predominance of flex-fuel and hybrid vehicle sales 50% flex-fuel and hybrid vehicles in the fleet Improvement in technology with a mix of flex-fuel and hybrid (greater combustion efficiency)	Predominance of flex-fuel vehicle sales More than 50% of flex-fuel and hybrid in the fleet Improvement in flex-fuel technology (greater combustion efficiency)

Table 17. Continued.

	'Pessimistic' scenario: countries reduce current targets. USA (blending targets 15% → 10%), China (blending targets 15% → 10%) and EU (10% → 5.75%)
6. Feedstock production capacity	Considerable rise in the world's population Lesser impact of climate change (1° C) No great improvements in seed technologies (fall of historical agricultural yield growth)
7. Biofuel production capacity	Machinery and equipment industries as barriers Stabilisation of the industrial facilities Big fall in growth rate of new units
8. Social improvement	Slave and child labour in developing countries Concentration of rural properties (large farms)
9. Environmental improvement	Failure of the Kyoto Protocol, difficulties in reaching new agreements, weakening of national and regional efforts towards reducing climate change New studies eliminate comparative advantages of biofuels from sugarcane and palm in terms of energy efficiency

'Expected' scenario: countries maintain current targets	'Optimistic' scenario: rise in current targets + adoption by other countries such as Russia and Japan
World population growth at historic rates.	Slow growth in world population
Maintenance of historic growth rates of agricultural yield.	Strong impact of climate change (3-5 °C)
Expected impact of climate change (3 °C)	Great improvement in seed technologies (GMOs, biofertilisers, more resistant varieties) with big advances in yield (beyond the impact of climate change)
Improvement in seed technology (technology able to match losses in yield due to climate change)	
End of obstacles related to base industry	Great technological improvement (viability of cellulose ethanol)
Maintenance of growth rates of new units	
Introduction of production through hydrolysis of cellulose and shared production (conventional technology + hydrolysis)	Rise in growth rate of new biofuel production units
No risks related to slave and child labour	Intense pressure from international organisations to redistribute agricultural income
Coexistence of the high-tech plantation model with the family agriculture integration models	Strengthening of agricultural contracts
	Total focus on family agriculture integration models
Countries meet the targets of the Kyoto Protocol, new agreements including developing countries (China, India and Brazil), regional agreements on emission control and successful climate exchanges in voluntary market	USA participation in global agreements on emission reductions, more ambitious targets in accordance with historical contribution, all countries adopting targets, according to contribution
Maintenance of comparative advantages of biofuels of sugarcane and palm in terms of energy efficiency	Improvement in the energy efficiency of all kinds of biofuels (sugarcane, palm, maize, beet, cassava, wheat, *Jatropha curcas*) and models of sustainable production and hydrolysis

Table 18. Summary of the strengths and weaknesses according to strategic areas in the sugarcane AGS (agribusiness system).

Innovation/ research/ production	Communication
Strengths	
Sugarcane has lower biofuel cost (than maize, beet, rapeseed)	Image of green fuel, jobs generator, environmentally correct, exporter, regional development promoter, and associated renewable fuel
Capacity of mature and large industry in Brazil	
Strong metal-mechanical industry dedicated to ethanol facilities	
Capacity of expansion to new regions in Brazil	'Free' advertising
Sugarcane varieties more resistant to climate change	UNICA (Sugarcane Industry Association)
Strong agronomic and biotechnological intelligence centres in Brazil	Communication campaigns in Brazil, USA and EU
'Genome Project' – mapping sugarcane genetic sequence	
Total use of by-products and residues in the field	
Flex-fuel technology	
Weaknesses	
Manual harvesting and human aspects in some sugarcane businesses	Low capability to anticipate problems and coordinate the response
Practice of burning (sugarcane)	Image of labour conditions during harvesting in developing countries
Low profitability of the sugarcane independent suppliers	Concentration of lands and farmers
Major investments in cellulosic ethanol research by the developed countries	Low corporate governance practices by the sugar mill sector
Big investments in hybrid car technology by the developed countries	

3.6 Objectives (goals) of the plan

The goals must be clear and consistent and, wherever possible, quantifiable. Thus, for the sugarcane chain, size-related goals must be thought through with regard to production and export volumes and others. These goals must also be seen in terms of economic sustainability

Distribution and logistics	Training and capabilities	Coordination and institutional environment
Vertical integration of ethanol facilities to distribution of fuels	Good training capacity (universities and research institutes) in Brazil	Consecana model (sugarcane payment formula)
Trading and oil companies' control of the sector	UDOP (Union of Bioenergy Producers) corporative university for executives and technical workforce	Agricultural partnerships
Bioelectricity's facilities concentration close to high demand for electricity and complementary to hydroelectricity sources		Associations and cooperatives
		Voluntary agreement to eliminate the practice of burning sugarcane
Export logistics in developing countries	Poor coordination between organisations that offer training (research institutes and universities)	Reputation/relational contract
Delay in ethanol pipeline infra-structure	Lack of executive and technical workforce	80-90% of the production cost of ethanol comes from sugarcane
Distribution cost	Reallocation of former sugarcane cutters	High vertical integration of biofuel facilities for agricultural production
Ethanol internal price fluctuation		Non-payment of sugarcane by fibre content
High concentration of fuel distributors		Lack of long-term contracts for distribution
Lack of fuel storage capacity in Brazil		Lack of pattern contracts for ethanol exports
Difficulty of connecting sugar mills to electricity grid		

(income of the main links in the production chain), environment (to keep the production bases for future generations), and people, aiming to include them in society, provide jobs, develop urban areas, etc. Table 19 provides some suggestions.

Table 19. Example of strategic objectives (goals) in 2015.

Type of goals	Description
Sugarcane production	Produce X tonnes at a target cost of US$ x and at a target price of US$ y
Ethanol production	Produce X billions of litres, representing 80% of Brazil's fuel consumption at a target cost of US$ x and a target price of US$ y
Energy production	Produce X MGW, representing 15% of Brazil's needs at a target cost of US$ x and a target price of US$ y
Sugar export	Export X tonnes to Y countries, representing 60% of world exports, at a target price of US$ x
Ethanol export	Export X tonnes to Y countries, representing 60% of world exports, at a target price of US$ x
Volumes of production units	Operating units
Profit margins in different links	Expected margins
Job volume	Expected jobs
Production of other products from sugarcane	Produce X litres of diesel and Y tonnes of plastic, among others

3.7 Main strategies

Brazil must pursue a cost-leading strategy with economic, environmental and social sustainability in order to supply all the market segments of sugar buyers, by type, country, needs and ethanol, besides many other generated products. Brazil's sugarcane industry should also position itself as one of the cleanest industries in the world, taking solar energy and transforming it into energy to be used by consumers.

To cope with these changes, we need to design, think strategically and change. The strategic projects that have been mentioned here reflect the authors' opinions, and they are meant only as impartial suggestions. Projects were defined for each of the areas (coordination and institutional adjustment, production and products, communication, distribution and logistics, and human resources) of a strategic planning process for the production system.

Of some help in preparing this material were the agendas already established by many important Brazilian agribusiness organisations such as UNICA[5] (Sugarcane Industry Association), UDOP[6] (Union of Bioenergy Producers), ORPLANA[7] (Sugarcane Growers

---

[5] http://english.unica.com.br/

[6] http://www.udop.com.br/

[7] http://www.orplana.com.br/

Association), CANAOESTE (Sugarcane Growers Association of Sao Paulo State), ABAG Ribeirao Preto[8] (Brazilian Agribusiness Association in Ribeirao Preto City), CTC[9] (Sugarcane Technological Center), IEA[10] (Agricultural Economics Institute) and IAC[11] (Campinas Agronomic Institute). A visit to their sites is also recommended for the latest information. There are major projects for the strategic area that can ensure the permanence of the current favourable conditions for the sugar-ethanol-energy sector. Of the projects presented in the next sections (3.8 to 3.12) some are exclusively from the private sector, others are public, and others are a combination of the two.

3.8 Projects and decisions related to production, products, research, development, and innovation

- Programmes for vertical growth of sugarcane production (higher yield in the same area) should be encouraged, through increased productivity, especially in genetic modification of sugarcane.
- Research and Development (R&D) integrated activities for the industry. Stimulation of the formation of public-private partnerships (PPPs) and technological parks between Embrapa, Agronomic Institutes, intelligence centres in universities, private companies, technological centres (such as CTC), and associations, with tax incentives and funds for the development of joint research in the sector. One approach to this proposal has come with the creation of the 'Brazilian Bioethanol Science and Technology Laboratory' by the government.
- Digital technology platform, which shows the current research, where and by which researchers it is being performed, promotes integration and helps prevent research and project duplication.
- Encouraging the integration and diversification of farming for food production and energy. Upon application of technologies to exploit the synergy between the two types of planting, demystifying the vision of competition among them. The integration of the sugar mill and ethanol distillery with biodiesel plant, livestock, maize, and other grains will add other products to the mix of mills. Dedini Co. has developed an integrated ethanol and biodiesel plant as an alternative to this movement.
- Strengthening a gene bank for sugarcane in order to meet the huge demand for new cane varieties that are resistant to pests and adapted to dry regions. The Sugarcane Technological Center (CTC) has been working in this way, and to continue this growth strategy they are moving towards becoming a public limited company (plc).

---

[8] http://www.abagrp.com.br/

[9] http://www.ctcanavieira.com.br/

[10] http://www.iea.sp.gov.br/

[11] http://www.iac.sp.gov.br/

- International patenting of technology and other aspects related to Brazilian ethanol production in order to avoid the 'free importation' of technology and to capture the value of technology export and thereafter of royalties.
- Product diversification by innovation so that the structures of production and industrialisation may be increasingly multifunctional. Today, sugar and ethanol facilities are taking advantage of economies of scope by adding the production of bioplastics and diesel.
- Permanent project for reduction of production costs in the production chain, aiming to capture greater value through lower costs, so improving profitability.
- Encouraging the expansion of cane activities, mainly in regions where there are degraded or underused pastures. This zoning should also consider that, for the local economy, it is important for other crops to remain, allowing coexistence with sugarcane and economic diversity. In this sense, one must adopt the zoning strategy of sugarcane production used by the Ministry of Agriculture (MAPA), by defining where it is permissible to plant sugarcane, and giving priority to the use of pastures, with no impact on the Amazon and the Pantanal Biomes.
- With regard to innovation, in an industry of large and small engines, it is necessary to study innovation through ethanol-powered motorcycles. Despite their low fuel consumption, they would contribute to improving air quality in major urban centres (a motorcycle pollutes 20% more than a new car). This innovation was presented by Honda Motors in 2009 and today represents more than 10% of the motorcycle market share.
- The adaptation of large diesel engines for ethanol with new technologies, aiming at the truck market for sugarcane suppliers and mills, as well as the market for tractors and urban buses. If mills could supply their own fleets of trucks with ethanol generated in-house, in a privileged tax regime, the cost could be reduced and might be passed on in the final price. Thus, this supply chain would be more environmentally friendly once the large volume of diesel consumed in operations is negatively summed on the emissions balance.
- Development of new products from the ethanol and sugar chemistry, besides those that have been developed, such as biodegradable plastic and diesel from sugarcane, etc.
- Innovations in production processes aiming toward improving the environmental balance of the activity.
- Projects for mill renovations, aiming at greater efficiency, including funding for new equipment such as more efficient boilers. Programmes to strengthen the base industry.
- Electric power production should be strengthened, to capitalise on existing potential in the sector, giving priority to this form of renewable energy through financing.
- Strengthen, in production, the ability to include small producers with sustainable remuneration, and the ability to establish and honour long-term contracts.
- Innovations for other products that could be processed at the mills, improving their use, and enabling them to be used for second- and third-generation ethanol.
- Innovations to improve industrial conversion processes, always looking for new sources of income.

3.9 Projects and decisions related to communication

In this area, many activities can be included in a plan for the sector. The main ones are highlighted below:

- Advertising an image of Brazil as a global supplier of renewable energy and environmental solutions. Strengthen the work of UNICA (Sugarcane Industry Association) and APEX (Governmental Agency for Brazilian Exports' Promotion) with the 'Agora' project to promote the image of Brazilian ethanol as 'sustainable fuel' – reduces countries' dependence on imported and scarce oil; encourage the adoption of clean technologies (flex-fuel, gasohol, local production in a sustainable manner, expansion of distribution network); ensure a sustainable production system, with high-energy balance (reduce emissions of greenhouse gases); allow the co-generation of clean energy (with the use of sugarcane bagasse), generating carbon credits.
- In partnership with municipalities and businesses, city buses could be tested on ethanol (UNICA has been doing this with the Agora project) on a much larger scale. In Sweden, there are 600, with only 3% above diesel cost. These buses would be painted and decorated with the production chain, and as such people would gain knowledge and information, either from the inside walls of the buses, or from distributed brochures. With a weekly changing of routes, in 4-5 months they would have 'spoken' to the entire user population of that county. Apart from the improvement in terms of air pollution in large cities, it would be a permanent channel of communication with the community.
- It is clear that with this continued growth it will not be long before 70 to 80% of fuel consumption in Brazil is ethanol. We will then have a petrol surplus. Petrobras will be able to export petrol ready for use, with added anhydrous ethanol, to neighbouring countries. It is a very real possibility that Petrobras could become the first green oil company in the world, but it has to work faster. Petrobras has a very important role to play in the image of ethanol, and ethanol (as well as biodiesel) has a very important role in establishing Petrobras' image. The Petrobras Group created a business unit dedicated to biofuels called 'Petrobras Biocombustivel' in 2008. They are investing in 'greenfield' biodiesel plants and in Acucar Guarani, the fifth largest ethanol company in Brazil, in partnership with Tereos International, a French commodities trading company.
- Create a list of priority countries for trade agreements (FTAs and tariff reductions) related to sugar and ethanol, and strengthen efforts in these countries. As examples, we have strong markets in Africa and Latin America for sales of sugar and ethanol as well as plant technology (facilities).
- Work on the development of African countries to jointly build an image of ethanol as a renewable, peaceful fuel.
- The gas station must be used directly as a communication tool for ethanol – 'green' stations, where the supply chain communicates with the final consumer. The sector has missed decades of opportunity to use this 'factory store' as a communication tool.
- Knowledge portal for sugarcane, which may be the UNICA website or other sources, which show everything that researchers and consumers need to know about sugarcane,

including databases of theses and dissertations, articles, books, and videos. It should be noted that this is the 'new media' generation, and therefore information should be offered to people by the appropriate means.

- We should also think about the green stations in California that sell E85 with sugarcane ethanol, promoting environmental balance and communicating even more with the American consumer. UNICA has promoted Brazilian sugarcane ethanol in USA gas stations as a way of contesting the import tariff of US$ 0.54/gallon.

3.10 Projects and decisions related to distribution and logistics

Among the decisions related to distribution and logistics, we can highlight the following:

- Mechanisms to encourage strategic stocks of ethanol to avoid price fluctuations, which harm the image of the product in the eyes of the consumer. Regulator stocks in Brazil and the main consumer markets of Brazilian ethanol can improve the image of the sector in Brazil and around the world and provide security of supply in domestic and international markets. In 2010, the ANP (Brazil's National Petroleum Agency) authorised the sale of ethanol between mills, stimulating the formation of stocks and commercialisation of bigger volumes throughout the year.

- The mills, with their focus on industrial activity, are very conservative about taking positions in distribution channels. Mills that are in the same region can build joint ventures to enter the ethanol distribution market, with an independent management, purchasing existing distributors or setting up new ones; these should be authorised to operate by the government (ANP). Any remaining arguments against this economic perspective today, are becoming increasingly weak as the volume of ethanol consumed in Brazil grows and may come to represent 70% of the market in five years.

- Also in the distribution channels, the mills in associative organisational forms, like franchises or joint ventures, could have gas stations in cities. These stations would be 'concept' stations (the name could be 'green' or 'verde'), and they would serve two basic functions: (1) to establish retail prices of ethanol (hindered by the action of urban cartels or the power of oil distributors), and (2) image communication for the final consumer, as stated in the item on communication (Section 3.9), because these stations would be green and decorated with the sugarcane and communication materials – in short, 'eco' service stations. The stations could sell petrol and diesel, but have ethanol in 80% of the pumps. The concept would be that of a 'factory store'.

- Improvement in infrastructure for distribution of ethanol. It is necessary to speed up already-announced investments in ethanol pipelines, as well as in port facilities for ethanol export at the lowest possible cost.

- Streamline the public-private partnerships (PPPs) and strengthen a broad privatisation programme of highways, railways, and ports in order to avoid overburdening the ethanol that comes from more distant regions, currently suffering from the problems of transport costs in Brazil.

- Strengthen alliances and joint ventures of ethanol-exporting groups to share investments and reduce risk, as well as joint activities in the logistics of national and international ports, freight, and other areas.
- Permanent innovation in harvesting and transportation logistics.
- General adoption of the standard contract for ethanol developed by IETHA (Association for International Trade of Ethanol), an organisation with almost 50 members. Before that, technicians from Brazil, EU, and the USA should work to standardise the fuel and transform it into a commodity.
- Once Brazil begins to dominate 60% or more of the world sugar market, companies should consider collective actions that will strengthen the logistics of transportation, port storage, and distribution of this product, with the aim of having very competitive costs.
- Easy access for mills to transmission lines (electricity grids) of the SIN (National Interconnected System), to enable them to strengthen the energy supply.
- International logistics of ethanol.

3.11 Projects and decisions related to training

Several actions are recommended in a plan for the sector in the area of human resources and training. It is noteworthy that with the growth of the sector, there is lack of people in this area. The Union of Bioenergy Producers (UDOP) has done excellent work in this domain by encouraging executives and technical professionals to act in the new sugar and ethanol mills.

The main suggestions are as follows:
- Map specific needs and coordinate the efforts of the existing organisations in the training of technicians and executives for the sugarcane production chain.
- Map the essential technical and undergraduate courses for the sugarcane agribusiness and its spatial distribution. Plan, along with many different organisations and the Ministry of Education, the granting of scholarships and incentives for research.
- Implement training programmes for workers (sugarcane cutters) in the sugarcane agribusiness system, organised by trade associations and unions. Implement training programmes for workers who have lost their jobs after harvest mechanisation. UNICA has been doing this with the 'Renovacao' project.
- Promote training of public employees related to agribusiness, in order to improve performance in the management of food quality, sustainability, certification, and traceability.
- Training initiatives in environmental sustainability in mills and farms.
- Establish a digital platform for training, aimed at popularising existing knowledge.
- Create a kind of 'Sustainable Regional Development' programme to stimulate sugar and ethanol mills, thinking about the inclusion of local communities. Propose corporate social-responsibility projects to add local companies and/or small producers in the sugar/ethanol mill supply chain. UNICA, together with SEBRAE (Brazilian Service of Support for Micro and Small Companies), could coordinate this kind of activity.

3.12 Projects and decisions related to coordination and adequacy of the institutional environment

Among the main activities in the area of coordination and institutional environment, we can highlight the following:

- First, the main points that federal and state governments could tackle are the tax and regulation issues. Ethanol has a reduced VAT rate of 12% in all states and a slight reduction in other federal taxes. Once virtually all the states are producing sugarcane, this reduction in revenues will be offset in part by production, the investments that have been and are being made, generated wages, and collected taxes. This does not take into account the environmental benefits and internalisation development.
- Study whether the addition of anhydrous ethanol to petrol could be expanded from the current range of 20-25% to almost 30%. Many people with petrol cars have already made this transition themselves. Thus, in cases of large production, the use of 30% could help consumption without any harm to the vehicle, and *vice versa*, thus contributing more to the environmental issue.
- In the case of new cars, it is necessary to think of instituting greater tax benefits for 'flex-fuel' vehicles as opposed to petrol-fuelled ones. The Brazilian domestic market is large, and there are market signals in favour of such a move (it is becoming increasingly difficult to sell petrol used cars in Sao Paulo, and soon this will be the case throughout Brazil). It has been observed that some car manufacturers such as Korean, German, Japanese, and American are still resistant, and thus Brazilian consumers have no access to larger cars with flex-fuel engines. It is also estimated that a large number of cheap vehicles will come to Brazil from China and India, and we cannot risk that they provide only a petrol option. American, French, and even Japanese manufacturers have proved that these flex-fuel engines are completely feasible. They could also, like the French manufacturers, export these cars and engines, spreading this technology and consumption to other markets.
- Still on the government agenda, it is necessary to consider changes in the tax burden for mills and distributors. In spite of the undeniable efficiency of fuel distributors, and their struggle to fight the informal market, better tax distribution through the value chain and states would greatly contribute to a more competitive market. ANP is working hard to increase the market efficiency through new rules and laws.
- The government should consider how to reduce labour costs in the industry, so that resources can be used in training programmes.
- Establish a unifying vertical trade association for the sector that would represent all the links in the chain. Promote planning, along with the government, through a strong and representative vertical trade association.
- The vertical association should be responsible for implementing marketing programmes and promoting sugar and ethanol exports (with a presence at major trade fairs and exhibitions abroad, in joint cooperation with Apex Brazil), with government and private funding.

- In the Consecana (sugarcane payment formula by sugar content) review, which is usually performed every five years, greater importance could be given to sugarcane bagasse (payment based on fibre content in the sugarcane) in the equation. In the future, with a more vertical system of production and distribution, Consecana can start from the final sugar prices on domestic and international markets and national average prices of ethanol at the pumps of gas stations.
- Define the certification process of Brazilian ethanol from sugarcane, coordinated by UNICA, to fit the industry standards for quality demanded by developed countries, mainly on the issue of sustainability. UNICA has been doing this by taking advantage of different types of social-environmental certifications like BSI (Better Sugarcane Initiative), Roundtable on Sustainable Biofuels, Ethos Institute, GRI (Global Reporting Initiative) and others.
- Stimulate the sector's dormant capacity to provide electricity through clear legal frameworks and purchase warrants, and preferential treatment for this type of energy, perhaps the source with least environmental impact.

Above are just some ideas from the authors' work in more than 10 projects in the sector. They have been out in the open for a long time, and some have already been implemented, either by existing organisations – and here we highlight the role played by the management of UNICA in terms of leadership, modernity, and globalisation – or by government and other bodies. Our proposal is that the coordination of this planning effort should be centralised, and that this could be accomplished by aiming at the sustainability of the sector, in order to make Brazil increasingly competitive and put it in the comfortable position of supplying energy to the world. In an age when the world needs water, food, and energy, sugarcane is, without a doubt, Brazil's best response to help meet these needs.

3.13 Conclusions

This chapter started with a macro-environmental analysis, which is very relevant for strategic planning and management in a production chain so as to align the agents´ efforts. Many agribusiness systems in Brazil attract international attention, but the sugarcane sector is different. Because of its history, the benefits it brings to Brazil and the world, its global leadership, and the internalisation of development to distant regions, it has acquired new defenders and policy makers in the last few years. At the same time, however, many concerns have emerged, especially as regards the sustainability of the production process.

The competitiveness of Brazilian ethanol is well recognised around the world. Its maintenance, however, depends on the operationalisation of important strategic projects, with the definition of responsible agents, deadlines, and budgets. More than ever, planning is necessary in this sector in order to take advantage of all the opportunities and to improve the weaknesses as the industry looks for equilibrium and sustainability. In short, this chapter tried to debate the main ideas surrounding 'the sugarcane road to improvement'.

4. An overview of FDI in the sugar-energy sector in Brazil and lessons learned

4.1 Characterisation of the largest groups in the sugar-energy sector

Ernst & Young's 'Biofuels Country Attractiveness Indices' (2008) ranks Brazil as the leading country in which to invest in both ethanol and biodiesel within the first two quarters of 2008. The index, which ranks the most attractive global markets for investments in biofuels, takes into consideration the existing infrastructure for biofuels (market regulatory risk, supporting infrastructure and access to finance) and some fuel-specific issues (off-take incentives, tax climate, grants and soft loans, size of new projects, currently installed base, domestic market growth, export potential and feedstock related issues, such as energy yield, sustainability and price volatility).

According to the report, Brazil leads the globalisation of the ethanol market for the recent inward foreign investment flows and its growing export capacity. Besides expanding production, the Brazilian ethanol industry (local and transnational companies) has been investing in overseas distribution assets, such as dehydration plants located in the Caribbean.

The other important player in the ethanol market is the USA. The Americans are not only the largest consumers but also the largest producers of ethanol. Nevertheless, Brazil's sugarcane ethanol has many advantages when compared to USA maize-based biofuel regarding energy yield, costs, environmental impact and competition with food. Unlike the USA, Brazil has plenty of favourable soil and climate available for producing feedstock. In addition, land prices in those areas where the sector is expanding are exceptionally attractive, labour costs are low and the country possesses the best technology within the sector.

The opportunities for exposure to the growing ethanol market have been attracting a number of new economic actors to the Brazilian sugarcane agribusiness system and the milling industry in particular. They come from other agricultural and industry-related activities, as well as from the financial market. The majority of the transnational companies that have made direct investments in the sector are large commodity trading companies and investment funds, but there are also some conglomerates and a few representatives of the automobile industry, the oil industry, the energy industry, the transport industry and the chemical industry, among others.

Most investments have flowed to the milling industry and the production of ethanol, power and sugar, but there have also been investments in R&D and infrastructure. According to the Brazilian Central Bank, the sector received around R$ 6.5 billion in the first three months of 2007 alone (Zanatta, 2007).

Its fragmented structure is another reason why the Brazilian milling industry creates opportunities for investors, although this picture has started to change. In the 2008/2009 crop year the 10 largest players in the industry were responsible for 40% of the overall sugarcane production. This makes the sugar and ethanol sector one of the most fragmented among Brazil's exporting industries. However, in 2005/2006 the share of the 10 largest groups was 30% of the sugarcane crushed in the country.

The four largest groups in the industry have been leading the consolidation with major merger and acquisition operations in the last couple of years. In October 2009 the French transnational food trader Louis Dreyfus Commodities (LDC) acquired a 60% equity stake in Santelisa Vale (SEV); this US$ 5.1 billion deal formed a giant with a total crush capacity of 36 million tonnes of sugarcane per year – until the crop year of 2008/2009 SEV was the second largest group with 5 mills that together could crush 16.7 million tonnes yearly. In December that same year, Bunge purchased all seven of Moema's mills for US$ 1.4 billion. In February 2010 Brazil's top producer, Cosan, announced a US$ 12 billion joint venture with Shell. The deal involves a total sugarcane crush capacity of 60 million tonnes, 4,500 fuel stations and the distribution of nearly 17.5 billion litres of fuel. One year earlier the company had purchased Nova America (6.8 million tonne capacity) and had entered the fuel distribution business by acquiring Esso's Brazilian subsidiary. Also in February 2010 ETH Bioenergia and Brenco merged to form a 9-mill company. Currently, the group can crush 10 million tonnes, but once all mills are operating at full capacity (2012), they will crush 40 million tonnes (Figure 22).

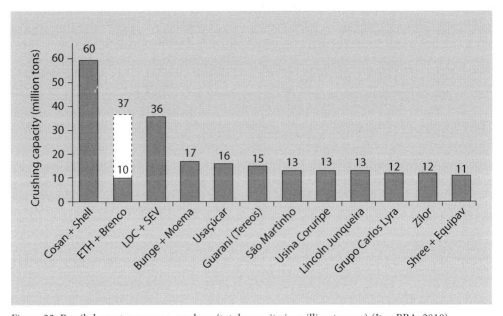

Figure 22. Brazils largest sugarcane crushers (total capacity in million tonnes) (Itau BBA, 2010).

Food and fuel

Until recently, the Brazilian sugar and ethanol industry was characterised by being almost completely financed by national capital. In the 2008/2009 crop year there were only two transnational companies among the main groups, Tereos (Açúcar Guarani) and Louis Dreyfus Commodities (LDC), both French. One year later, Bunge, Shree Renuka Sugars and Shell have joined them in the group of companies whose capacities exceed 10 million tonnes per year. Others, like ETH+Brenco, have minor but important shares of foreign capital. Table 20 summarises the foreign player's position in the sugarcane industry.

The majority of their plants are located in the state of Sao Paulo, where 61% of Brazil's sugarcane is grown. In the 2008/2009 harvest, more than 50% of sugarcane expansion took place in Sao Paulo. However, the largest growth rates are to be found in Minas Gerais, Goias and Mato Grosso do Sul, indicating that these are the new frontiers for sugarcane expansion.

4.2 The institutional and regulatory framework

The arrival of FDI in Brazilian agribusiness intensified as deregulation took place in the mid 1990's. In 1996, activities related to agriculture and livestock attracted US$ 568 million, or 6% of total FDI in the country in that year. In 2006, FDI in these activities was more than 500% higher than a decade earlier, reaching US$ 3.5 billion and 16% of total inward FDI. These figures relate to the purchase of assets and greenfield ventures (Chiara, 2007). Today, a TNC may own an industry, as well as farm operations, via a subsidiary company based in Brazil.

Table 20. Overview of the foreign players in the sugar and ethanol industry (based on company reports).

Foreign players	Sector of origin	Country of origin	2009/2010 crushing capacity (million tonnes)
Cosan + Shell	Oil	The Netherlands	60
LDC + SEV	Trading	France	36
Bunge + Moema	Trading	USA	17
Tereos (Guarani)	Sugar	France	15
Shree Renuka + Equipav	Sugar	India	11
Infinity Bioenergia	Fund	USA	8
Abengoa (Dedini)	Energy	Spain	7
CNAA	Fund	USA	5
ADM	Trading	USA	4
Adecoagro	Fund	Argentina/USA	4
Noble Group	Trading	Hong Kong (China)	3
British Petroleum	Oil	UK	2
Cargill	Trading	USA	1.3

In the milling industry, the movement in mergers and acquisitions started in the early 2000s, but was mainly restricted to local companies. Back then the scenario was similar to that which the mills are currently facing. The industry was fragmented, mostly owned by families, and suffered from huge debts and stagnation. However, within the country there was some tough competition and the milling products were becoming more valuable, starting with the sugar prices in 2000 and 2001, and later the ethanol prices upon the arrival of flex-fuel cars in 2003. The most capitalised local companies took the chance to purchase the devalued assets of their competitors with the purpose of expanding and reinforcing their position. Along with gains in scale and lower administrative costs, centralisation led to a somewhat more professional approach towards administration and enabled better access to financing. This process naturally appealed to domestic and foreign investors and also increased the number of mills.

The first transnational corporation to enter the Brazilian milling industry was the French Louis Dreyfus Commodities (LDC), which in 2000 acquired two mills in the state of São Paulo and one Minas Gerais. In 2007, Louis Dreyfus became one of the largest companies in the sector after purchasing the group Tavares de Melo with its five plants. According to Bruno Melcher, president of the administrative council of LDC Bioenergy, the eight plants owned by the company crushed 15 million tonnes of sugarcane during the 2008/2009 harvest and expect to increase production to 20 million tonnes in 2009/2010 (Scaramuzzo, 2008).

Tereos was the other transnational to enter Brazil's milling industry before the biofuel boom. The group, which consists of 13 French cooperatives and 12,000 associates, took control of Açúcar Guarani in 2003 after incorporating the French group Béqhin-Say, which had acquired Guarani's five mills in 2001. To assure liquidity and expand its sugarcane production throughout 2009, Açúcar Guarani announced a plan to raise between R$ 193 and 309 million through private subscription (Valor Economico, 26/01/2009, 27/01/2009 and 17/02/2009).

More recently, with the ethanol market reaching international proportions, the role of foreigner players in M&A has grown significantly, as has the number of overall operations. In 2003 there were five recorded operations, while in 2007 there were 25, of which 70% of transactions were conducted by investment funds and large corporations. Foreign capital was responsible for 12% of all sugarcane crushed in Brazil in the crop year of 2004/2005. Today, this share has already reached 27%.

Such a boom in investment inflated the price of plants. The average price of Brazilian mills reached R$ 110.00 per tonne of sugarcane crushing capacity in early 2008, more than double the average recorded in 2005, which was R$ 45.00. With the outbreak of the credit crisis, the total number of M&A fell to 14 in 2008. In 2009 the prices of assets have already fallen. The current average is around R$ 85.00 per tonne of sugarcane crushing capacity.

In a fragmented industry, the financial crisis speeded up consolidation in 2009. The shortage of credit complicates the operations and there have been more mergers than acquisitions. Nevertheless, the devaluation of assets creates opportunities for capitalised large foreign corporations. Investors usually purchase or join up with existing mills (brownfield investment) since greenfield investments can take up to four years depending on the size and development of the sugarcane crops.

UNICA has focused most of its efforts on trying to enlarge the international market for Brazil's ethanol. The association has especially invested in promoting the product's image abroad, supporting sustainability policies and practices among its associates, reducing tariffs and other barriers to trade in developed countries, and also attracting local and foreigner investors.

Table 21 summarises the foreign players' growth strategies in the Sugarcane Industry.

4.3 The growth strategies in the sugar-energy sector

Table 22 presents a synthesis of the four major new businesses, highlighting the production processes, the market potential, the advantages that these new businesses provide to mills, and the main companies which are involved or have been investing in these segments. It is true to say that TNC groups are looking for investment opportunities throughout the sugarcane value chain.

An interesting trend in the sector is the entrance of companies from other sectors looking for diversification and value capture. These companies usually join forces with traditional groups within the sugarcane sector or even create new groups. Table 23 shows a clear picture of the sector of origin, investment locations, and the main objective of the main new players in the sector.

Oil companies will be looking increasingly at carbon emissions, and entering into the sugarcane ethanol business must have two points of attraction compared to other fuels: scale and compatibility with the existing infrastructure. British Petroleum announced that biofuels were chosen because of energy security, climate change, and rural development. The company believes that biofuels will account for 20% of the market in 2020. BP and Dupont are also developing a technology, to be launched in 2013, to produce biobutanol from sugarcane, which has 20% more energy capacity than ethanol.

Shell's deal with Cosan forms a giant worth US$ 12 billion that encompasses almost the entire chain, from sugarcane production to fuel pumps. Petrobras (Brazil's largest oil company) has created Petrobras bioenergy with the goal of becoming one of the five largest fuel companies in the world. Petrobras sees sugarcane-based ethanol as an important fuel for the future, in the local market and as an oxygenate for international markets. Its investments in biofuels

Table 21. Foreign players' growth strategy – organic growth vs. M&A.

Group	Recent M&A	Recent greenfields
Cosan	Benálcool (SP); Nova América (2 mills); Esso Brazil (distribution); Shell (distribution)	Acreúna, Jataí and Rio Verde (GO)
Santelisa Vale	Vale do Rosário (SP)	CNAA (4 mills in MG)
LDC	Tavares de Mello (4 mills); Santelisa Vale (5 mills)	Rio Brilhante (MS)
Tereos (Guarani)	Andrade and São José (SP); Vertente (SP)	Tanabi (SP), Cardoso (SP)
São Martinho	-	Boavista (GO)
Usina Coruripe	-	Carneirinho, União de Minas, Prata (MG)
Alto Alegre	-	Santo Inácio (PR)
Usaçúcar	-	Terra Rica (PR)
Grupo Moema	-	Frutal and Bom Jardim (MG), Indiaporã (SP)
Bunge	Grupo Moema (4 mills); Santa Juliana (MG)	3 mills (TO); Nova Ponte (MG)
Infinity Bioenergia	Disa (ES), Paraíso (MG), Ibirálcool (BA)	Jateí, Laranjal, Iguatemi I and II(MS)
Bertin	Infinity Bioenergia (3 mills)	-
Clean Energy Brazil	Unialco (MS); Usaciga (PR)	Santa Mônica (PR); Rio Paraná (MS); Pantanal (MS); Água Limpa (GO)
ETH Bioenergia	Alcídia (SP); Eldorado (MS); Brenco (4 mills)	Conquista do Pontal (SP); Caçu (GO)
Brenco	ETH Bioenergia (5 mills)	Ati Taquari (MT), 3 mills (GO), Costa Rica (MS)
Abengoa	Dedini (2 mills in SP)	-
Shree Renuka Sugars	Equipav (2 mills in SP)	-
Noble Group	Petribú Paulista (SP)	One mill in SP
British Petroleum	-	Tropical Bioenergia (GO)

are expected to reach US$ 2.4 billion from 2009-2012. Petrobras also believes that in 2020, 75% of the Brazilian car fleet will be flex-fuel, 17% petrol, 1% full ethanol and 7% diesel.

4.4 The role of investment funds (private equity)

After deregulation took place in the mid-1990s, the role of the State as main promoter and financer of the sector shrank. Recently, although there are different forms of governmental support, the expansion of the sector has been based on the approach taken by the mills and financial groups. One of the new sources of finance that has attracted Brazilian mills is the local stock market. Currently, Cosan, São Martinho and Açúcar Guarani are listed on the São Paulo Stock Market. Many other have announced their IPOs in the next few years. This movement towards the financial system offers opportunities that local and foreigner private equity funds have not ignored.

The operation of private equity funds has played an important role in the expansion of Brazil's capital market. In the milling industry, these funds have intensified their business, especially since 2006. The financial assets they administrate come mainly from institutional investors, such as pension funds, government funds and large corporations. They are led by venture capital managers who prepare the companies to receive investments and maximise their value. Usually, after entering the business, the managers of private equity strategically intervene in the organisational structure of the companies, frequently replacing employees at senior levels. The goal is to create value to the business, through financial engineering and improvements in management, for the moment of disinvestment, i.e. when the fund withdraws its assets and makes its profits. The most common way of disinvestment is through an IPO. The period between investment and disinvestment can last a couple of months or many years, depending on the nature of the business and the pace of its valorisation.

The first investment fund to make direct investments in the Brazilian milling industry was Infinity Bioenergy in 2006. The group, which is listed on the London Stock Market and has Merrill Lynch, Stark and Och-Ziff Management among its investors, entered the country through the American fund Kidd & Company, which was later incorporated by Infinity. At the time of the entry, the group announced that it had some US\$ 800 million to invest. Infinity Bioenergy initiated its entry by acquiring five plants in the states of Minas Gerais, Espírito Santo and Mato Grosso South (Valor Economico, 17/09/2007). Nevertheless, liquidity problems drove the company to negotiate 71% of its shares with Bertin, originally a beef producer and processer that has grown in the infrastructure, concession and energy businesses.

At the end of 2006 another investment group was founded looking to profit from the Brazilian milling industry. The Clean Energy Bio-Energy (CEB), which is also listed on the London Stock Market, is formed by Brazilian and European investors and rose 200 million pounds at its IPO. Its goal is to crush 30 million tonnes of sugarcane per year.

Coupling firms with private equity funds has been a major strategy helping local corporations to expand. In 2006, Brazil's second largest sugar and ethanol producer, Santelisa Vale, became partners with the Sugar and Alcohol Investment and Participation Fund (FIP) which is formed by Global Food, Carlyle and Riverstone, Goldman Sachs and Discovery Capital. Together, they invested in a major merger operation with Vale do Rosário, one of Brazil's largest mills.

Table 22. Main new businesses for the sugar-energy sector (based on interviews, Valor Econômico, 2009 and O Estado de S. Paulo, 2009).

Product	Production process	Market potential
Plastic	Ethanol production Dehydration Polymerisation	Viable with oil price under US$ 45/barrel 4 million tonnes of green polyethylene in 2015 In 2012 Brazilian projects will produce 810 million tonnes of green plastics which is equivalent to 1.95 billion litres of ethanol Total market of 68 million tonnes of polyethylene (2007) 9% of all oil produced goes to plastic production Plastic production is a business generating US$ 350 billion per year
Hydrolyses/ cellulosic ethanol	Acid (standard)	1/3 of the cane's biomass is found on the straws and leaves
	Enzymatic (more efficient) Cracking of the cellulose molecule into sugar, which later ferments producing ethanol	140 million tonnes of bagasse (2007/2008) 1 billion litres of ethanol (10% bagasse residue which is not burned) 14 litres/tonne of cane by 2015 37 litres/tonne of cane by 2025
Bioelectricity	High pressure and temperature boilers Bagasse burning Straw burning (trend) Research for transforming bagasse into pellets	Each tonne of sugarcane produces 250 kilos of bagasse Current joint capacity of mills reaches 5,300 MW 3,000 MW are added to the network (3% of energy matrix) In the south-central region there are 210 ventures with total capacity of 11,500 MW by 2015 Using all straw and bagasse, by 2020 and 2021, the potential of the sector would reach 14,400 MW
Fuel distribution	Pulverised ethanol market – 420 mills Concentrated fuel market – 198 fuel distributors	In 2008 106 billion litres of fuel were traded on the local market, ethanol being responsible for 18% of this volume Revenue of R$ 186 billion in 2008 Over 50% of light duty vehicles run on ethanol
Diesel	Uses genetically modified yeasts to ferment the sugar present in the cane and produces diesel	Internal diesel demand reaches nearly 45 billion litres of diesel yearly, of which 5 billion are imported The target is to produce 400 million litres in 2011 and 1 billion litre in 2012 Is economically viable with oil price under US$ 60/barrel

Advantages	Companies
Fluctuation of oil prices	Braskem
Renewability	Dow
Low water costs	Cargill
Recyclable	Du Pont
	Solvay
Higher yield, increasing from 50% to100%	Novozymes
Better use of resources, reducing the use of land	DuPont and Genencor
Opportunity cost of the use of bagasse for generating	Dedini
bioelectricity	CTC
Reduction of 50% in greenhouse gases	Shell
Use of non-food raw material	Petrobras
	Amyris
Complements the hydroelectricity	All mills for self-supply (400 units)
Proximity to consumers	Some mills sell the surpluses (210 units)
Low investment needs	Only one uses the straw for energy (Equipav)
Reduced construction time	
Low environmental impact	
Pulverised energy generation	
Sale of carbon credits	
Energy self-sufficiency for mills	
Learning about the operations	Market share: Petrobras (19.3%), Shell (12.6%),
Transference price control	Ipiranga (12.1%), Chevron (7.2%) and Esseo (5%)
Proximity to consumers	Cosan bought ESSO and merged with Shell Brasil
Stock integration and transport optimization	Grupo Ultra bought Ipiranga Sul, Sudeste and
Purchasing guarantee	Texaco (Chevron)
Increase in negotiation power with other distributors	Petrobras bought Ipiranga Norte, Nordeste and
Likely taxation advantages	Centro-Oeste
Possibility of acquiring value in the purchase with	
small mills	
Produces no sulphur content	Amyris + Sao Martinho
Renewable source	Monsanto
Can be mixed with common diesel in a proportion	LS9 + Jalles Machado
of up to 80%	
Has the same chemical characteristics of	
conventional diesel	

Table 23. New players in the sugarcane sector and their strategies (based on interviews, Valo Econômico (2009) and Unica (2009).

Company	Region of activity
Investment groups and funds	
Infinity Bioenergia (American and Brazilian Capital)	Controls mills in Bahia, Minas Gerais, Espítiro Santo and Mato Grosso do Sul
Cluster de Bioenergia (Brazilian)	Plants in Mato Grosso, Mato Grosso do Sul, Goiás, Minas Gerais and Bahia
Goiás Agroenergia (90% American Capital)	New production units in Goiás
Agroindustrial groups	
Multigrain: JV between PMG Trading (Brazilian), Mitsui (Japanese) and a CHS (American).	Plant, under construction in the west of Bahia
Grupo Samam (Brazilian)	Construction of another plant in Sergipe
International Paper (American)	Long-term contract for the purchase of bioenergy from Equipav
LDC (Louis Dreyfus Commodities) (French)	Plants in the south-eastern, central and north-eastern regions
Campo Lindo (Brazilian)	New industrial complex in Sergipe.
ADM (Archer Daniels Midland) (American)	New industrial complex in Jataí, Goiás
Tradings	
Noble Group (Chinese)	New sugar and ethanol mill in São Paulo
	Construction of a terminal at the port of Santos, in the state of São Paulo
Toyota Tsusho (Japanese)	Construction of a plant in the southwest of Goiás
Cargill (American)	Participation in plant Bom Jardim, with Moema in Itapagipe, Minas Gerais
	Expansion and diversification of plant Cevasa (São Paulo), controlled by the group since 2006

Strategic actions and plans	Strategic goal
Greenfield and brownfield	Attractive return on investment rates
Sold 71% of its shares to Bertin (originally from the beef industry)	Portfolio diversification
Investment of $ 3 billion	
By 2015 all units should be operating and crushing capacity is set to reach 10 million tonnes of cane sugar	
The goal is to have 30 thousand hectares of sugarcane	Return rates
Mitsui, which holds 25% of the shares, closed a deal with Petrobras and the Camargo Correa (construction) for building a pipeline for ethanol exports	
One plant in operation in Sergipe	
The supplier plant is a (large scale) pioneer in cogeneration from straw and tips	
One of the first multinationals to enter the business, in 2000	
Investments of around US$ 6 billion, mostly in acquisitions	
Will produce 700 thousand litres of ethanol per day and generate 30 MW of energy	
Initial crushing capacity of three million tonnes	
Purchase in 2007, the Petribú Paulista Mill, located in the interior of São Paulo	Return rates
	Supply guarantee
Partnership with Petrobras and cane producers in the region	
Investment of more than US$ 500 million	

Chapter 4

Table 23. Continued.

Company	Region of activity
Tradings (continued)	
Bunge (American)	Plants in the South-eastern and Central regions
Construction companies	
Odebrecht – ETH Bioenergia S/A (Brazilian)	Construction of at least nine new sugar and ethanol plants in Mato Grosso do Sul, Goiás and São Paulo
	Construction of plant in Angola
	Plans for plants in Latin America and Africa
Oil companies	
Petrobras (Brazilian)	Creation of Petrobras Biofuels
	Is seeking a partnership with USA oil company for a project to produce ethanol in Brazil and exported to the USA
	A distillery of ethanol in the Midwest, with the Japanese Mitsui
	Investing in projects to build ethanol pipelines
	Cooperation agreements with the Italian Eni in the areas of biofuels and heavy oil
	Research for production of second-generation ethanol
Royal Dutch Shell (the Netherlands)	-
British Petroleum (UK)	Has a 50% holding in Tropical Bioenergia, a plant in Goiá, in a JV with Maeda and Santelisa
Chemicals	
DuPont (American)	Technology for production of second-generation ethanol in the USA
Lanxess (German)	Construction of an energy co-generation plant using bagasse from sugarcane to supply its plant in Porto Feliz (Sao Paulo)

An overview of FDI in the sugar-energy sector in Brazil and lessons learned

Strategic actions and plans	Strategic goal
Alliance with the Japanese ITOCHU for the development of projects. The first joint venture with the Japanese group involves the Santa Juliana mill in Brazil Purchased Moema Par (100% of 2 mills and 60% of shares in 2 others)	Return rates Supply guarantee
Merged with Brenco. Plans IPO for 2011 Japanese trading Sojitz owns 33% of the group Full capacity of 40 million tonnes in 2011/2012	Business diversification Investment in infrastructure
The goal is to be a leader in the national production of biodiesel and expand the participation in the business of ethanol, with focus on the international market Establishment of Brazilian ethanol, in partnership with Nippon Alcohol Hanbai who will be responsible for marketing Brazilian ethanol to Japan Eni is to develop biofuel projects with Petrobras around the world Intends to build at least 15 ethanol plants by 2012 in partnership with major companies	Ensuring the supply of additive Concept green/renewable Survival New position – energy
JV with Cosan for production and distribution of 1st and 2nd generation ethanol, bioenergy and sugar (US$ 21 billion revenues) Together, control 4,470 fuel stations	New position – energy
Investing in 2nd generation ethanol and looking for new partners for greenfield projects Was the first oil company to enter the business	New position – energy
Initial investments of US$ 140 million	Alternative to oil supply
With the project, the company expects to reach almost zero CO_2 emissions from the unit, which today is around 44 tonnes per year	Technology

Table 23. Continued.

Company	Region of activity
Chemicals (continued)	
Braskem	Is a branch of Odebrecht, which has shares in ETH Bioenergia
Dow	Had started a JV project of US$ 1 billion with Santelisa Vale
Solvay	No investments in ethanol plants
Energy	
Cerona (Companhia de Energia Renovável) (American)	Plant projects in the cities of New Andradina and Batayporã (Mato Grosso do Sul)
CPFL (Brazilian)	Acquisition of bioenergy in São Paulo Partnership with Baldin Brothers to co-generation CPFL Bioenergy
Rede Energia (Brazilian)	Contract for buying energy from bagasse from Cosan
Tractebel Energia (French/Belgium)	Partnership with Açúcar Guarani in the generation and sale of energy from bagasse from sugarcane
Banks	
Morgan Stanley (American)	The Bank is defining where allocate their investments in Brazil, but must be near ports, such as Santos (São Paulo)
Technology	
Celltrion (Biotechnologies) (Korean)	One plant in Bahia
Monsanto (American)	Purchase of Alellyx and CanaVialis (R&D)

The partnership also formed a new company called Companhia Nacional de Açúcar e Álcool (CNAA). This JV received a total of R$ 2.2 billion from these partners in order to build four greenfield projects in the states of Minas Gerais and Goiás. Together, they will be able to crush 21.6 million tonnes. Two of them are already operating and a third was set to start running in the second semester of 2009. Currently, each of these three units can process 2.7 million tonnes of sugarcane annually, but the company announced last February that it

An overview of FDI in the sugar-energy sector in Brazil and lessons learned

Strategic actions and plans	Strategic goal
Its sugarcane-based polypropylene is the first certified 'green plastic'	Alternative to oil supply
Not on commercial scale yet	
After the merger of Santelisa with LDC the project is on stand-by	Alternative to oil supply
In February 2010 announced investments of US$ 500 million in the production of sugarcane ethanol based PVC	Technology
Initial milling capacity of about 10 million tonnes of cane per harvest	Construction of the cogeneration plant and transmission lines
The plants will also receive investments in co-generation of energy from sugarcane bagasse	Guaranteed Supply
Bought R$ 500 million in bioenergy from Cosan	Complementarity to hydropower
The value of the contract with Cosan was estimated at R$ 489 million	
The French multinational Areva Koblitz will build a thermoelectric plant in Sao Paulo, which will use the bagasse as fuel	
Before the crisis, the bank's commodities area expected to be responsible in 2008 for 10% to 15% of Brazilian exports of ethanol	Return rates
Investments of R$ 500 million	Green concept
Companies in the sugarcane sector towards development of new varieties of cane	

would be investing US$ 418 million in order to expand their joint capacity to 16.2 million tonnes. Around US$ 275 million were invested by Riverstone, while the Inter-American Development Bank poured in the other US$ 143 million. CNAA has also announced the decoupling from Santelisa Vale, which held 25% of its share. CNAA is now solely controlled by the Sugar and Alcohol Investment and Participation Fund.

Although the Brazilian milling industry is remarkably fragmented, foreigner investment groups found it difficult to acquire stakes in traditional family owned mills before the crisis, given the valorised plants and the resistance of owners to share power within their businesses. Therefore, until the beginning of 2008, foreigner investors had to invest in greenfield projects in regions where sugarcane production was not traditional. This also shows their long-term perspective in the sector.

Investments funds are also taking the long-term position in the sector. Brenco (Brazilian Renewable Energy Company) started from a pool of local and foreigner investors willing to enter the milling industry and became one of the largest ethanol producers and exporters. The group is backed by Henri Reichstul (former president of Petrobras), Vinod Khosla (a co-founder of Sun Microsystems), Bill Clinton (former president of the United States), James Wolfensohn (former president of the World Bank and former director of the National Petroleum Agency), Amber Capital, Ashmore Energy International, the Yucaipa Companies, Tarpon Investments, and others.

Brenco's R$ 5.5 billion investments have been directed to greenfield projects in country's Middle West region, the main sugarcane expansion area. All of the sugarcane harvesting will be mechanised in order to avoid environmental and social criticism about manual harvesting. The group has become a reference on how to structure the business in order to obtain investments. It has adopted a vertically integrated approach to the sugarcane chain believing it helps provide access to the international market. According to the company, a vertically integrated organisation helps reduce costs and control quality as well as technical, environmental and social specifications.

Another fund that has deployed long-term direct investments in the milling industry is Adecoagro. The fund, which is formed by American and Argentinean investors, has the major investor George Soros as its main partner. Adecoagro bought the Usina Monte Alegre, Minas Gerais, in 2006, and is currently building a new plant in Mato Grosso do Sul. The company plans to invest R$ 1.6 billion to reach a processing capacity of 11 million tonnes of cane by 2015.

As Mundo Neto (2008) explains, the financial system demands new, more transparent organisational structures than those typically seen in family businesses, where power is concentrated in the hands of family representatives, usually the patriarch. In the Brazilian financial system, the most common way to professionalise management and get exposure to financial investors is via corporate governance. This kind of structure, where the decision-making power is held by an executive board and an administrative council, tends to reduce conflict between managers and shareholders.

These changes have affected not only the management and organisational structure of the mills, but also the technical procedures both in the industry and in the sugarcane fields. The use of technology is being intensified with the objective of maximising production efficiency and undermining criticism regarding sustainability. In the plantations, the mechanisation of all operations is clearly a trend. In those mills that have not yet mechanised the harvesting, working conditions have been improved considerably. In fact these trends are not the consequences of the operation of foreigner capital, but they certainly accelerate the process.

Most private equity funds have been seeking exposure to the blooming ethanol market by investing in the industrial side of the business – which usually also involves the agricultural side, due to the vertically integrated structure of the sector. Nevertheless some of these funds have found an opportunity to profit by focusing on the agricultural side only.

That is the case with Brookfield Asset Management Incorporation, a global investment group based in Canada, which has been operating in Brazil since the early 1900s. In Brazil it operates through its local subsidiary, the Brascan Group, investing in agricultural land, timberland, real estate, infrastructure and power generation. As far as agriculture is concerned, Brascan has around 150 thousand ha in Brazil, on which it grows grains and sugarcane, and raises cattle. Its sugarcane area currently accounts for 17 thousand ha. Therefore, sugarcane is part of the company's broader strategy for agriculture and real estate. This strategy consists of acquiring agricultural areas which they believe have a great potential for valorisation in the long term as well as performing profitable activities in the short term. Those are mostly ranches where cattle are raised on an extensive basis. After acquisition, improvements are made to the property and herd in order to valorise the assets. From that point on, Brascan's activities in those areas might shift according to its return expectations, commonly migrating from livestock to agriculture. Most recently, the opportunities present in the sugarcane business have led the company to purchase land where the crops are being expanded. The company then looks for investors that are willing to invest in greenfield projects for a mill. However, it does not take part in the industrial entrepreneurship. Once the deal is done, Brascan either leases the land to the company that owns the mill or grows the sugarcane and delivers the crop to the plant afterwards. That way, profits are made according to the momentum of agriculture as well as from the return on capital invested in the land. From the point of view of the crushing industry, such models allow investments without incurring capital immobilisation in land and with a more flexible governance structure. That way, for instance, newcomers with no expertise in growing sugarcane would be better to purchase the feedstock produced by Brascan.

4.5 Financing (leverage debt equity)

Brazil has already proved to be one of the most competitive countries in agribusiness. Nevertheless, despite being one of the most important producers and traders of agricultural products, there is much to be improved and the country still lags far behind its true potential as a food and energy supplier to the rest of the world. Brazilian farmers, agri-food industries, traders, services suppliers, government leaders and financial institutions face growing demands for quantity, quality, sanity and sustainability, among other things.

These market demands require new forms of administration, production processes, distribution and regulation. Since these changes cost money to implement, Brazilian agriculture and agribusiness is looking for new sources of capital in order to overcome the State's lack of resources for the sector and the local credit market's high interest rates and fees. Inward FDI is clearly playing a major role in financing the expansion and consolidation of the Brazilian Sugarcane Agribusiness System.

Transnational sugar and ethanol trading companies also have a part to play in financing mills and their sugarcane production. One of the most common ways for mills to finance their production is through prepayment. They sell their products, especially sugar, ahead of the next harvest to trading companies in order to generate working capital. Unlike grains, many of the sugarcane products are traded by national companies such as Copersucar, although transnational trading companies are becoming more relevant in financing sugarcane production, albeit not (yet) on the same level as for soya production, for instance.

The State has also played an important role in financing the expansion of the sector, especially through the National Bank for Economic and Social Development (BNDES). The bank finances investments in the industrial production of ethanol, sugar and electricity, the agricultural production of sugarcane as well as in infrastructure. The BNDES portfolio of active operations in the milling sector totals US$ 21.3 billion. In 2008, the bank lent R$ 6.5 billion to projects for planting sugarcane, producing ethanol and sugar, and initiatives in the co-generation of electricity by power plants. In the last 5 years, this amount has increased by US$ 13.8 billion. With the crisis the importance of BNDES for the sector has grown even bigger.

Some other projects have emerged after the foundation of the group, such as the Petrobras pipeline, which will link the Middle West region and Brazil's biggest port, passing through some of the most important production areas.

4.6 Corporate social responsibility

The sugar industry in Brazil is very developed in terms of corporate social responsibility. Among the major groups that make up part of the UNICA Industry Association, these practices are linked to the sustainable development of people. UNICA and its member companies continually develop programmes aimed at improving labour conditions and establishing national benchmarks.

According to the national Annual Social Information Report (2008, cited in Moraes, 2009) this industry is one of Brazil's most relevant in terms of job creation – around 1.3 million jobs. Research conducted by UNICA showed that the average wage paid by member companies was double the current federal minimum wage. Brazilian laws comply with International Labour Organisation standards covering working conditions and are subject to frequent government inspections. The cane cutters have collective labour agreements, and innovative programmes to improve labour conditions are being put in place, including the elimination of outsourcing for manual sugarcane cutters, better transportation standards, and increased transparency in performance measurements and employee compensation. These conditions have to be met by local investors as well as by TNC companies.

UNICA also has a socio-environmental responsibility indicator that tracks corporate responsibility performance in the industry, with the aim of encouraging best environmental and sustainable practices. Other projects include the Social Balance Program developed with the Brazilian Institute for Social and Economic Analysis (iBase) and data gathering for UNICA's Global Reporting Initiatives on Sustainability (GRI). In 2008, member companies invested over R$ 160 million in 618 projects within social, environmental, cultural, education, sport and health areas, benefiting approximately 480,000 people.

There are several lessons to be learned from this experience. One of the most important is that if a particular sector that is receiving a TNC investment has a strong association, and this association is managed according to the most recent international standards and in an environmentally responsible way, the possible negative impacts of an FDI and the risks of exploiting people and human resources will almost certainly be mitigated. The example of UNICA (the Brazilian Sugarcane Industry Association) followed by other organisations of farmers, is of fundamental importance and an important lesson for TNC investments.

4.7 Environmental regulations and trends

One of the most relevant aspects that could be learned on this subject is the activities of UNICA, who, together with the Brazilian government and the State Secretary are working on environmental regulations. The major targets of UNICA are:
1. *Fewer agrochemicals.* The use of pesticides in Brazilian sugarcane fields is low and the use of fungicides is practically non-existent. Major diseases that threaten sugarcane are fought with biological controls and advanced genetic enhancement programmes that help identify the most resistant varieties of sugarcane. Thanks to the innovative use of recycled production residues such as vinasse and filter cake as organic fertilisers, Brazilian sugarcane production uses fewer industrialised fertilisers than other major crops.
2. *Low soil loss.* Brazilian sugarcane fields have relatively low levels of soil loss, thanks to the semi-perennial nature of sugarcane, which only needs replanting every five to six years. Future trends indicate that current losses, however limited, will decrease significantly in coming years through the use of sugarcane straw, some of which is left on the fields as organic matter (mulch) after mechanical harvesting.

3. *Minimal water use.* Brazilian sugarcane fields require practically no irrigation because rainfall is abundant and reliable, particularly in the main south-central production region. Rainfall is complemented by fertirrigation, a process that involves applying vinasse, a water-based residue from sugar and ethanol production which is rich in organic nutrients. Water use during industrial processing has decreased significantly over the years, from around 5 m^3 per tonne to approximately 1.5 m^3 per tonne of sugarcane processed. With improved technologies such as dry wash, the industry expects to reduce water use further in the coming years (based on UNICA data).

According to UNICA, one of the most important initiatives launched by the sugar and ethanol sector is the 'Green Protocol', also called the 'Agro-environmental Protocol', signed in 2007 between the sugarcane industry and the São Paulo State government. Under the agreement, the industry agreed to speed up the elimination of sugarcane burning, a traditional practice that facilitates manual cane harvesting.

The 'Green Protocol' brought forward from 2021 to 2014 the eradication date for areas where mechanised harvesting is currently possible, and from 2031 to 2017 the deadline for other areas, for example steep slopes. The Protocol also states that as of November 2007, new sugarcane fields must have fully mechanised harvesting. By mid-2009, some 157 sugar and ethanol plants in São Paulo State had signed the Green Protocol (80% of all companies in the state). 24 sugarcane producers' associations signed the protocol, which represents 4,745 growers and 587,462 ha of cultivated lands. Mechanised harvesting already responds to 50% of the sugarcane production in the state. Sugar and ethanol producers, together with labour organisations and different levels of government, are developing job training and requalification programmes to mitigate the effects of mechanisation on sugarcane cutters.

With the Protocol being followed by industries, producers and government, more than 227 thousand hectares of original forest are being recovered in Sao Paulo State. The consumption of water has also been reduced to an average of 1.92 m^3 per tonne of cane crushed. It was more than 5 m^3 a few years ago, and 53% of the industrial units in Sao Paulo are already using less than 1 m^3.

5. Concluding remarks – FDI: suggestions for regulations, public policies, and incentives

5.1 How to evaluate the capacity of international investment to generate the regional development

This chapter has been developed using the authors' experience with projects that attracted international investments for regional development, and the experiences of the suppliers, farmers and other stakeholders. The field research and interviews conducted for this study, examining the impact of FDI on sugarcane business in Brazil, also contributed to the analysis. In short, we will discuss the importance of receiving international investments for all nations, but mostly for developing nations, and we will try to bridge the 'gap' in the analysis.

Local or federal governments and other institutions sometimes experience difficulties evaluating the capacity of an international investment to promote economic development. This difficulty can make it hard for governments to define a specific benefit or support to an international investment company, and even to convince the local community of the benefits for developing the region, growing the economy, generating jobs, exports, etc.

When a transnational company (TNC) comes to a country, it normally comes with several types of resources, not only financial. These resources are our basic list for analysis. We have categorised these resources in a list of six points, in order to help governments to evaluate international investments in the country. The investments are better if an investor performs well on this list of resources. Our list is different, since it also looks at the local supply chain of the company. If a company can build a good integrated supply chain, there is a better chance of more economic development. Each of the following six 6 topics has its own sub-topics. It might help to understand the relevance of the points by imagining an international food industry that is investing in a new country.

1. *Financial investments and expertise.* Here we may consider the amount of money that will be invested and linked to this point, if this company can provide capital, can open credit lines giving the necessary guarantees for suppliers (for instance, local farmers), can have access to government official credit, has knowledge of credit operations and bureaucracy, has access to international credit and generates a good reputation for the region and the country. We should also look at the capacity of generating the benefits of foreign currency through the increase in exports and, finally, the amount of employment generated.
2. *Capacity to provide technical assistance.* Normally, a TNC has a 'how-to-do package' for its suppliers, helping with farm support, support for sustainability policies and sustainable practices, participating in R&D activities, helping to achieve standards (ISO, etc.) and transferring skills that will promote economic development.

3. *Sourcing of input supplies to farmers.* A TNC company can help farmers by providing them with modern seeds, machinery, genetic know-how, fertilisers and chemicals, thus helping them to produce with the most recent technology.

4. *Management assistance and service provisions.* Here we should evaluate the capacity of assistance with economical/financial controls for farmers and suppliers, training and farming management, transportation and storage, communication skills and certification. Supporting local communities in their demands for public investments in regional logistics and infrastructure is relevant as well.

5. *Capacity to provide market access.* This is one of the most important points. A TNC can arrange international sales contracts, providing access to marketing channels and niche markets – organic/fair trade/others -, providing information on market trends, helping farmers decide what to grow, and enabling a reduction in price volatility through long-term contracts.

6. *Farmers' and suppliers' organisations.* This final point is not usually acted on or evaluated. We think a TNC should also be evaluated by its capacity to help farmers or suppliers build what is called 'countervailing power', to reduce power imbalances in modern food production chains, although it may appear nonsensical. This can be done by stimulating the establishment of local organisations, stimulating the setting up of cooperatives, building networks of local producers and incentives for cooperation.

The list of major 6 points to help governments evaluate TNC investments is summarised in Table 24.

5.2 How to promote and regulate international investments

Brazil is a country that is generally very flexible when it comes to receiving international investments, and for that reason attracts a large amount of investment by TNCs. As discussed before, TNC companies can own land and industry in all production chains. The institutional and regulatory framework concerning TNC participation (e.g. FDI, PPP, contract farming) in the production of sugarcane, is an interesting collection of legislation from a wide array of ministries. Some of this legislation is very old and does not differ for a national or transnational investor. Moreover, there isn't a specific range of policies, but rather an array of different policies at the national, regional or local level.

There is plenty of information about the possible benefits of international investments in an economy. In this section, we will show that a country wanting to receive these investments should set up an institutional arrangement, since these international investments may have positive and negative impacts. With this in mind, some regulations are important to try to prevent the negative impacts of transnational companies' investments, and to enhance the positive effects of these direct foreign investments.

Table 24. Impact of TNCs on regional development.

Resources of TNCs	Impacts on regional development
Financial investments and expertise	Providing investment (capital)
	Opening credit lines giving the necessary guarantees to farmers
	Access to government official credit
	Knowledge of credit operations and bureaucracy
	Access to international credit
	Inward investments contribute to a good reputation for the region and the country
	Foreign currency generation through the increase in exports
	Employment generation
Input supply to farmers	Seeds
	Machinery
	Genetic know-how
	Fertilisers and chemicals
Technical assistance	'How-to-do package'
	On-farm support
	Support for sustainability policies and sustainable practices
	Execution of research and development
	Support for standards (ISO, etc.)
	Transferring skills
	Supply of higher standard products within the internal market
Management assistance and service provisions	Assistance with economical/financial controls
	Training and farming management
	Transportation and storage
	Communication
	Certification
	Sales/income of the farm
	Support for demands for public investments in logistics and infrastructure
Market access	Arranging sales contracts
	Providing access to marketing channels
	Access to niche markets – organic/fair trade/others
	Providing information on market trends, helping farmers decide what to grow, and enabling a reduction in price volatility
Farmers' organisation	Establishment of local organisations
	Stimulating the setting up of cooperatives
	Network of local producers
	Incentives for cooperation

There are 8 major topics that should be studied and covered by public policies. The objective here is to facilitate local, state or even federal governments and agencies in setting up a framework which attracts international investments to promote development and prevents possible negative externalities.

The first topic relates to the governance structure of the investments. In this area, we can look at the wide array of possible investments (joint ventures, vertical integration, franchises, etc.), money entrance conditions, the promotion policies to receive these investments, safeguards for risk protection (invasion, expropriation, fees, etc.) and others. How will direct investment take place and what sorts of asset ownership (land, industry and others) must be considered and planned? How should the stimulus package for these investments (like energy supply, logistics and other related to infrastructure) be built and how must the existing obstacles to investments be removed?

The second topic relates to environmental protection, focusing on policies of water usage, agricultural practices (soil preservation and harvesting, among others), policies on pollution control, sanitary measures, international standards and certifications that will be required and, finally, policies regarding the preservation and rights of biodiversity. Some companies are accused of not having the same environmental practices as they have back home, a situation that should be prevented by means of suitable policies.

The third topic deals with the regulation of human resources. These regulations may include salaries, labour and wages, benefits, working conditions, corporate social responsibility, ethics and codes of conduct and community relations. This is one of the most important topics, since most problems concerning past international investments occurred in the area of bad human resource management.

Taxation policies are the fourth topic that must be defined for transnational investments. Questions regarding the structure of taxes and tax policies, export tax policies, purchase and compensation taxes and possible government temporary tax incentives for the investment to be made or to be stimulated are the focus of the analysis here.

In fifth place, we have research and development policies. At this point, the most relevant would be a kind of stimulus package to improve the development of local knowledge and R&D. Property rights, licensing contracts and royalties must be discussed. Stimulus packages for linkages with local research organisations and institutions may be an important incentive to integrate and promote development.

The sixth topic is more related to agricultural or agribusiness investments, and deals with joint activities for farmers and industry. It is important to have policies that stimulate the linkage between international investments and local organisations, an incentive for cooperatives and association's formation and sustainability, previous preparation of farmers, co-ops or organisations for relationships with the international investments and incentives for building sustainable supply contracts. It might also be important to establish a framework for dispute mechanisms and even private arbitration.

The seventh topic concerns financing and credit: discussing and implementing policies on how an international investor can access public sources of financing, state banks and public credit lines. This offer of credit, linked to the technology of the international investor, offers a great opportunity for growth.

Finally, the last topic relates to policies regarding market access. In each of these policies, there are suggestions for incentives for international investments. These may include government purchasing of products generated by the investment and facilitating local access to investors, international agreements for market access to improve export channels of this new entrant and general competition policies. In food investments, it is also important to evaluate and promote food safety policies to facilitate international market access.

The list of 8 major points to help governments to regulate TNC investments are summarised in Table 25.

Table 25. Recommended public policies for the regulation of foreign investments in agribusiness.

8 major topics	Recommended public policies, incentives for international investments in agribusiness
Governance structure	How direct investment will take place and type of asset ownership (land, industry and others)
	Entrance conditions of resources (money flows)
	What kind of promotion policies for FDI
	What kind of safeguards for protection against risks (invasion, expropriation, fees, etc.)
	What kind of stimulus package for investments (energy, logistics and others related to infrastructure) and how to remove the obstacles to attracting investments
Environmental protection	Policies on water use
	Policies on agricultural practices (soil preservation, harvesting, among others)
	Required international standards and certifications
	Policies on pollution control
	Sanitation policies
	Policies on the preservation and rights of biodiversity

Table 25. Continued.

8 major topics	Recommended public policies, incentives for international investments in agribusiness
Human resources (people and labour)	Rural labour & wages
	Working conditions
	Benefits
	Community relations
	Child labour
	Corporate social responsibility
	Ethics and codes of conduct
	International labour
Taxation	Structure and tax policies
	Export and tax policies
	Purchase and compensation taxes
	Temporary tax incentives
R&D (science and technology)	Development of local knowledge and incentives for local development of R&D
	Property rights and other protection forms, licensing contracts and royalties
	Linkages with local organisations/institutions as an incentive
Joint activities for farmers and suppliers under contract (commitment)	Linkages to local organisations
	Incentive for co-ops/associations formation and sustainability
	Prepare farmers/co-ops/organisations for the relationships
	Sustainable supply contracts
	Dispute mechanisms and private arbitration
Financing and public resources	Access to public sources of financing
	Access to state banks and credit lines
Policies on market access	Government purchasing policies and access to investors
	International agreements for market access
	General competition policies
	Food safety policies for market access

5.3 A strategy for international investments

There is still a debate about whether or not these investments are good for a particular country or region. We should not advocate for one side or the other, but collect all the arguments and then make the analysis. International investments do have positive points, if they promote development, by bringing access to international markets and expanding the country's capacity for exports, creating jobs and generating taxes for governments, bringing knowledge to a country, bringing credit, and giving confidence, among other things, to a country. If a world class company invests in a country, it is an endorsement for development and other investors.

The major objections to international investments are linked to expatriation of resources from countries, exploitation of these resources till they are used up so preventing future generations of that country from using them, sending profits away from local economies and bringing cultural shock and cultural change to local communities. Others fear the damage to competition in a country, due to the global capacity of a multi-national company that can even promote dumping on local markets, compensated by good results in other countries, in order to destroy local competition. This may cause the exclusion of local companies in the long term. There are also some nationalist feelings, that only products produced in a local country, by a local company, can be good for the local community. These points should be considered.

But in our view, with a good strategy and a good regulation system, a country can try to avoid the downsides, maximising the benefits of these investments. The first requirement, which is fairly uncommon among national governments, is to have a good national strategic plan, looking 10-20 years ahead. With a good strategy it is possible to attract companies to a country or region linked to the potentialities of that region, with expertise, guaranteed demand (international contracts), clean production systems, and high technology (biotech/nanotech). But also a guarantee that units of research and development as well as parts of the headquarters of these companies would be established in the country that is receiving this investment.

Brazil is an example of a country that is experiencing huge flows of international investment, major economic growth and income distribution, but this is putting pressure on its infrastructure and booming internal markets. Brazil will also be hosting two major events that companies could take advantage of: the World Cup in 2014 and the Olympic Games in 2016. This is comparable to the situation of China, which experienced similar growth in investments (Beijing, 2008). A good strategy is to attract international investment and companies that are in harmony with the conditions of the country and its opportunities. For example, Spanish, American, Asian and other hospitality chains (like Melia, Hilton, Sheraton, Shangri-La) who could introduce and expand their networks of hotels (business); investments in entertainment (parks, museums); Chinese businesses wanting to produce energy and infrastructures (roads, trains, airports); airlines (now allowed to own 49% of local companies' shares) which are among the fastest growing sectors; construction investors, building second homes for retired Europeans on the north-eastern beaches of Brazil (six hour flight from Europe); and investment in universities, a booming sector due to the demand for education. There is also room for investment in food, for instance New Zealand dairy farmers wanting to expand globally, Belgian chocolate companies, and

Australian/Uruguay sheep farmers and slaughterhouses. These are just a few areas where a strategy would fit.

When approaching international markets to attract investments, all countries and their respective governments should do the groundwork, or at least look at the basics. These include ensuring that the economy is strong (growth, low inflation, favourable interest rates, internal demand), having a supply of good human resources and talent, a reasonable and competitive infrastructure, security, reasonable tax and financial systems, straightforward and corruption-free administration to facilitate management. A country also has to offer basic resources (energy, land, sun, water), good suppliers and distributors and institutions (judiciary system) that are trustworthy and can rapidly solve problems and disputes. This groundwork, together with a good strategy, good regulation systems and simplicity, will create an environment for international investment to promote sustainable development.

5.4 Planning strategies for 2010-2020

To conclude, we would like to point to the following 10 important strategic areas for the next ten years, ranging from strategic planning processes to the overall positioning strategies of companies, consumers and governments. We hope the list will help shape readers' thoughts and actions for the period 2010-2020.

1. *Empowerment.* In the next ten years, companies, networks and production chains will be more valued by consumers. How to build sustainable incentives for associations and cooperatives will be a central task for governments.
2. *Economic integration.* A more integrated economy increases the importance of the developing countries' supply chains, as sources/alternatives for supplying consumers in the developed world. This integration also brings prominence to the marketing channels in the developing countries through which products made in the developed world could be sold to emerging consumers.
3. *Income distribution.* There is a huge internal market growing in several parts of the world, and these emerging consumers should be studied more by companies for their positioning. The impact of these new consumers on the planet's capacity to produce (from grains to proteins) is a major issue.
4. *Climate and environment (preservation).* This topic will be even more important in the next 10 years, since climate change is a reality. The following are essential: attention to low carbon networks (carbon footprint and management), network adaptation to climate change, renewable energy networks, environmental certification, resource usage efficiency, network reversal (material reuse and recycling) and network integration for optimisation of usage of by-products.
5. *Technology.* As the drivers of cost reduction, consumers will not only value network transparency, information exchange and technology systems, but also consumer 'hi-

touch' networks. Companies and their networks should really be driven by consumers and communicate with consumers on an individual basis.

6. *Industry mergers.* The next 10 years will be significant for product mergers, which have already taken place in the area of mobile phones – now used as cameras, computers, watches, voice recorders, and radios – and other equipment. The world will see the growth of networks in nutraceuticals (food and pharmaceutical), nutri-cosmetics (food and cosmetics), nutri-tourism (food and tourism business), and nutri-cars (food and biofuels).

7. *Risk management.* An integrated network of risk management and mitigation will be of fundamental importance in this connected world. Several risks are present in a global perspective, such as financial crisis, disease, sustainability, and security.

8. *Communication.* There will be crucial changes in this area, with new media network communication, proactive network communication with stakeholders. Communicating with high-tech consumers will be a challenge for companies.

9. *Era of simplicity.* Simplicity will be valued, in terms of company network management, market segmentation, new product launching, brand management, services, costumer focus, sales management and others. Simplicity is the keyword for the next 10 years!

10. *Network value engineering.* This is an era of integrated company networking, which includes the promotion of permanent supply chain redesign, the value capture of marketing channels, customer intimacy, contract evaluations and reliable relationship building.

5.5 Final discussion

Finally, the relevance of studies in TNC involvement in agribusiness and agriculture is of fundamental importance. Food production needs to be enhanced for several reasons. Population growth, economic growth and income distribution, urbanisation, use of food crops for biofuels production, and other major trends create a scenario in which food production needs to be doubled in the next 20 years. TNCs are vital in this area

At a time when countries are establishing policies on food security, with governmental and private funds being allocated to purchasing land abroad and securing food supply, the role of TNCs is becoming increasingly important.

The sector chosen for this study in Brazil is the fastest growing in the country's economy. It is a sector in which the sustainable production of ethanol has been taking place for more than 30 years. It is an industry where TNC investments from different sectors are happening each day. While this report was being produced, the second largest Brazilian group was transferred to a French multinational company. It is a large sector which will double in size in the next 15 years, receiving investments from all over the globe.

There are several lessons to be learned by developing nations trying to attract international investments, either in agriculture, industry or even starting up sugarcane operations to generate electricity, sugar and ethanol. This study showed the major groups entering the sector, the basics of the industry and how this industry and the country organised policies to receive these investments, as well as the importance of strong organisations and associations.

References and websites

ABDI, Agência Brasileira de Desenvolvimento Industrial and Núcleo de Economia Industrial e da Tecnologia do Instituto de Economia da Universidade Estadual de Campinas – UNICAMP (2008). Relatório de Acompanhamento Setorial. Transformados Plásticos, Vol. 2, July, 2008.

ANFAVEA, National Automotive Vehicle Manufacturers Association (2010). São Paulo, Brazil. Available at: http://www.anfavea.com.br/tabelas.html.

ANP, National Agency of Petroleum, Natural Gas and Biofuels (2010). Rio de Janeiro, Brazil Available at: http://www.anp.gov.br.

Aston, A. (2009). China's Green Innovation. Business Week China, May 1, 2009. Available at: http://www.iscvt.org/news/articles/article/?id=43.

Batalha, M.O. (ed.) (2001). Gestão agroindustrial. 2nd ed. São Paulo, Brazil: Atlas.

Bourlaug, N. (2007). Vocação da terra. Revista Agroanalysis 27(3). Available at: http://www.agroanalysis.com.br/index.php?area=conteudo&esp_id=10&from=especial&epc_id=80.

BP, British Petroleum (2006). Statistical Review of World Energy. London, UK: BP. Available at: http://www.bp.com.

Bradesco, Departamento de pesquisa e Estudos Econômicos (2009). Agronegócio em Análise. Available at: http://www.bradescorural.com.br/site/conteudo/analise/agronegocio.aspx.

Camargo, J.M. (2007). Relações de trabalho na agricultura paulista no período recente. Tese Doutorado em Ciências Econômicas, Instituto de Economia da Universidade Estadual de Campinas. Campinas, Brazil: Universidade Estadual de Campinas.

Campomar, M.C. and Ikeda, A.A. (2006). Planejamento de Marketing: e a Confecção de Planos. São Paulo, Brazil: Editora Saraiva.

Carvalho, E.P. (2003). Demanda externa de etanol. In: Seminário Álcool – Potencial de Divisas e Emprego. Rio de Janeiro, Brazil: BNDES. Available at: http://www.bndes.gov.br/conhecimento/publicacoes/catalogo/s _alcool.asp.

Carvalho, L.C.C. (2004) Setor Sucroalcooleiro. Seminar MBA Agronegócios Fundace/Pensa. Ribeirão Preto, 10 December 2004.

Central Bank of Brazil (undated). Available at: http://www.bcb.gov.br.

Chiara, M. (2007) Agroenergia atrai capital externo. O Estado de São Paulo. 29 January 2007.

EMBRAPA CPAA, Empresa Brasileira de Pesquisa Agropecuária Amazônia Ocidental (2007). Available at: http://www.cpaa.embrapa.br/.

Ernst & Young (2008). Biofuelds Country Attractiveness Indices. Renewable energy country attractiveness indices. Quarter 1-2, 2008. Available at: http://frankhaugwitz.com/doks/bio/Industry_Utilities_Biofuels_country_attractiveness_indices.pdf.

F.O. LICHT'S (2007). World Ethanol & Biofuels Reports. Kent, UK: Agra-net. Available at: http://www.agra-net.com.

FAO, Food and Agriculture Organization of the United Nations (2008a). FAOSTAT. Rome: Italy. Available at: http://faostat.fao.org/site/377/default.aspx.

FAO, Food and Agriculture Organization of the United Nations (2008b). Available at: http://www.fao.org/.

References and websites

FAO, Food and Agriculture Organization of the United Nations and OECD, Organization for Economic Cooperation and Development (2007). OECD-FAO Agricultural Outlook: 2007-2016. Available at: http://www.agri-outlook.org/dataoecd/55/42/39098268.pdf.

Financial Times (2009) Saudis Set Aside US$ 800 million for Foreign Food. Financial Times 14/04/09. Available at: www.ft.com.

Global Demographics (2008). Available at: http://global-dem.com/.

Goldman Sachs (undated). Available at: http://www2.goldmansachs.com/.

IBGE, Instituto Brasileiro de Estatística (2007). Available at: http://www.ibge.gov.br/.

ICONE, Institute for International Trade Negotiations (2007). São Paulo, Brazil. Available at: http://www.iconebrasil.org.br.

IEA, International Energy Agency (2004a). Biofuels for Transport: An International Perspective. Paris, France. Available at: http://www.iea.org/textbase/nppdf/free/2004/biofuels2004.pdf.

IEA, International Energy Agency (2004b). World Energy Outlook 2004. Paris, France Available at: http://www.worldenergyoutlook.org/2004.asp.

IEA, International Energy Agency (2005). World Energy Outlook 2005. Paris, France. Available at: http://www.worldenergyoutlook.org/2005.asp.

IEA, International Energy Agency (2006). World Energy Outlook 2006. Paris, France. Available at: http://www.iea.org/textbase/press/pressdetail.asp?PRESS_REL_ID=187.

IETHA, International Ethanol Trade Association (undated). Available at: http://www.ietha.org/ethanol/.

IMF, International Monetary Fund (undated). Available at: http://www.imf.org/external/index.htm.

Itau BBA (2010). Visão e tendências do setor de Açúcar e Álcool no Brasil. Available at: http://www.itaubba.com.br.

Jain, S.C. (2000). Marketing Planning & Strategy, 6th edition. Cincinnati, OH, USA: Thomson Learning.

Johnson, G. and Scholes, K. (1997). Exploring Corporate Strategy. 4th edition. Upper Saddle River, NJ, USA: Prentice Hall.

Klein, B., Crawford, R.G. and Alchian, A.A. (1978). Vertical Integration, Appropriable Rents, and the Competitive Contracting Process. The Journal of Law and Economics, Vol. 21(2), pp. 297-326.

Leal, M.R.L.V. (2006). O teor de energia da cana-de-açúcar. NIPE, Núcleo Interdisciplinar de Planejamento Estratégico. UNICAMP, Campinas, Brazil. In: F.O.Licht's 2nd Sugar and Ethanol Brazil. March 2006, São Paulo, Brazil. Available at: www.nipeunicamp.org.br.

Macedo, I.C. Seabra, J.E.A. and Silva, J.E.A.R. (2008) Greenhouse gases emissions in the production and use of ethanol from sugarcane in Brazil: The 2005/2006 averages and a prediction for 2020. Biomass and Bioenergy, Vol. 32 (4), pp. 582-595.

Mapa Ministério da Agricultura, Pecuária e Abastecimento (2010a). Projeções do Agronegócio: Brasil 2009/2010 a 2019/2020 / Ministério da Agricultura, Pecuária e Abastecimento. Assessoria de Gestão Estratégica. Brasília, Brazil: Mapa/ACS

Mapa, Ministério da Agricultura, Pecuária e Abastecimento (2010b). Ministério da Agricultura, Pecuária e Abastecimento. Intercâmbio comercial do agronegócio: principais mercados de destino / Ministério da Agricultura, Pecuária e Abastecimento. Secretaria de Relações Internacionais do Agronegócio. Brasília, Brazil: Mapa/ACS.

Mathews, J.A. (2008). Towards a Sustainably Certifiable Futures Contract for Biofuels. Energy Policy, Vol. 36 (5), pp. 1577-1583.

Mathews, J.A. and Goldsztein, H. (2007). Capturing Latecomer Advantages in the Argentinean Biofuels Industry. In: VI International PENSA Conference, Ribeirão Preto, Brazil, October 2007.

MB Agro (2007). Oferta e demanda de fertilizantes no Brasil: uma avaliação da dependência externa da agricultura brasileira. Associação Brasileira de Marketing Rural e Agronegócio. Available at: http://www.abmra.org.br/marketing/insumos/fertilizantes/oferta_demanda_fertilizantes_mbagro.pdf.

Moraes, M.A.F.D. (2000). A Desregulamentação do Setor Sucroalcooleiro do Brasil. Americana, Brazil: Caminho Editorial.

Moraes, M.A.F.D. (2009). Externalidades sociais dos diferentes combustiveis no Brasil. In: Seminario Etanol e Bioeletricidade: a contribuição da cana para o desenvolvimento sustentável, ESALQ, December 2009.

Neves, M.F. (2007a). A Method for Demand Driven Strategic Planning and Management for Food Chains (The ChainPlan Method). In: 17th Annual World Forum and Symposium – Food Culture: Tradition, Innovation and Trust – A Positive Force for Modern Agribusiness, Parma, Italy, June 2007.

Neves, M.F. (2007b). Strategic Marketing Plans and Collaborative Actions. Marketing Intelligence and Planning, Vol. 25 (2), pp. 175-192.

Neves, M.F. and Thomé e Castro, L. (2009). Inserting Small Holders Into Sustainable Value Chains. In: Handbook of Business Practices and Growth in Emerging Markets, S. Singh (ed.), Hackensack, NJ, USA: World Scientific Publishing Co., pp. 235-253.

Neves, M.F., Waack, R.S. and Marino, M.K. (1998). Competitividade no Agribusiness Brasileiro: Sistema Agroindustrial da cana-de-açúcar. Sao Paulo, Brazil: Pensa/Ipea.

Neves, M.F., Pinto, M.J.A. and Conejero, M.A. (2009). Transnational Companies Investments in Brazilian Agribusiness and Agriculture: The Case of Sugar Cane. In: UNCTAD (United Nations Conference on Trade and Development) World Investment Report 2009. Geneva, Switzerland: UNCTAD. Available at: http://www.unctad.org/Templates/Page.asp?intItemID=1465&lang=1.

Neves, M.F., Trombin, V.G. and Consoli, A.M. (2010). Mapping and Quantification of the Sugar-Energy Sector in Brazil. In: Proceedings of 2010 IAMA (International Food And Agribusiness Management Association) World Symposium & Forum, 2010, Boston, MA, USA.

NIPE/UNICAMP, Núcleo Interdisciplinar de Planejamento Energético (2005). Available at: http://www.nipeunicamp.org.br/.

NREL, National Renewable Energy Laboratory (2006). From Biomass to Biofuels: NREL Leads the Way. Golden, Co, USA: NREL. Available at: http://www.nrel.gov.

NYMEX, New York Mercantile Exchange (2007). Available at: http://www.nymex.com/index.aspx.

OECD, Organization for Economic Cooperation and Development (2005). Producer and Consumer Support Estimates, OECD Database 1986-2004. Paris, France. Available at: http://www.oecd.org/document/9/0,3343,en_2649_33773_35015433_1_1_1_1,00.html.

OICA, International Organization of Motor Vehicle Manufacturers (2007). Available at: http://oica.net.

Poschen, P. (2007). Green jobs and Global Warming. ILO (International Labour Office). Available at: http://www.ilo.org/global/About_the_ILO/Media_and_public_information/Feature_stories/lang--en/WCMS_087408/index.htm.

Probiodiesel, Programa Nacional de Produção e Uso do Biodiesel (2007). Available at: http://www.biodiesel.gov.br/.

RFA, Renewable Fuels Association (2008). Annual Industry Outlook. Renewable Fuels Association, Washington, DC, USA. Available at: http://www.ethanolrfa.org.

Rothkopf, G. (2007). A Blue Print for Green Energy in the Americas: Strategic Analysis of Opportunities for Brazil and the Hemisphere. Featuring: The Global Biofuels Outlook 2007. The Inter-American Development Bank.

Scaramuzzo, M. (2008) Grupo Louis Dreyfus faz planos para elevar produção de etanol. Valor Econômico, São Paulo, 18 December 2008. Available at: http://www.valoronline.com.br//ValorImpresso/MateriaImpresso.aspx?dtmateria=18/12/2008&codmateria=5326869&codcategoria=305.

Secex, Secretaria de Comércio Exterior do Ministério do Desenvolvimento, Indústria e Comércio (2009). Estatísticas em AliceWe. Available at: http://aliceweb.desenvolvimento.gov.br.

Steenblik, R. (2007). Biofuels – At What Cost? Government support for ethanol and biodiesel in selected OECD countries. The Global Subsidies Initiative of the International Institute for Sustainable Development (IISD), Geneva, Switzerland. Available at: http://www.globalsubsidies.org/IMG/pdf/biofuel_synthesis_report_26_9_07_master_2_.pdf.

UNICA, Sugarcane Industry Union (2010). Available at: http://www.portalunica.com.br/portalunicaenglish/?Secao=lectures%20and%20presentations.

USDA (2010). Production, supply and distribution. Available at: http://www.fas.usda.gov/psdonline/psdQuery.aspx.

USDA, Foreign Agricultural Service (2010). GAIN (Global Agriculture Information Network) Report Biofuels. Available at: http://www.fas.usda.gov.

Van Dam, J. Jungingera, M., Faaij, A., Jürgens, I., Best, G. and Fritsche, U. (2006). Overview of recent developments in sustainable biomass certification. Copernicus Institute for Sustainable Development, Utrecht University, and the Environment and Natural Resources Service, Food and Agriculture Organization of the United Nations, Rome, Italy, Oeko-Institut (Institute for Applied Ecology), Darmstadt, IEA Bioenergy Task 40. Available at: http://www.fairbiotrade.org_files\fwd.html.

WBCSD, World Business Council on Sustainable Development (2002). Move. Sustain. The Sustainability Mobility Project. Available at: http://www.wbcsd.org/.

WBCSD, World Business Council on Sustainable Development (2004). Mobility 2030: Meeting the challenges to sustainability. The Sustainable Mobility Project. Full report. Available at: http://www.wbcsd.org/web/publications/mobility/mobility-full.pdf.

Williamson, O.E. (1985). The Economic Institutions of Capitalism: Firms, Markets, Relational Contracting. New York, NY, USA: The Free Press, 449 pp.

Williamson, O.E. (1996). The Mechanisms of Governance. New York, NY, USA: Oxford University Press.

World Bank (2008). Available at: www.worldbank.org.

WWI, the Worldwatch Institute (2006). Biofuels for Transportation: Global Potential an Implications for Sustainable Agriculture and Energy in the 21st Century. Extended Summary, Washington, DC, USA. Available at: http://www.worldwatch.org/taxonomy/term/445.

Xinlian, L. (2009). In the Name of Green. Beijing Review, May 21, 2009, pp. 32-33.

Zarrilli, S. (2007). The emerging of biofuels market: regulatory, trade and development implications. UNCTAD (United Nations Conference on Trade and Development) BioFuels Initiative. New York, NY, USA and Geneva, Switzerland. Available at: http://www.unctad.org/en/docs/ditcted20064_en.pdf.

Zanatta, M. (2007) Etanol, febre movida a subsídios nos EUA. Valor Econômico, São Paulo, 6 November 2007. Available at: http://www.valoronline.com.br/ValorImpresso/MateriaImpresso.aspx?codmateria=4621680&dtmateria=2007-11-6&codcategoria=306.

Food and fuel

Zylbersztajn, D. (1995). Estruturas de Governança e Coordenação do Agribusiness: Uma Aplicação da Nova Economia das Instituições. Tese (Livre-Docência), Departamento de Administração, Faculdade de Economia, Administração e Contabilidade, Universidade de São Paulo, São Paulo, Brazil.

Zylbersztajn, D. and Neves, M.F. (eds.) (2000). Economia e Gestão dos Negócios Agroalimentares. São Paulo, Brazil: Pioneira.

Annex 1. Most relevant M&A operations from 2007 to 2010.

Effective date	Target company	Buyer	% of acquired shares	EV (US$ MM)	Transaction value (US$ MM)	Milling (MM tonnes)	EV/milling (US$/ton)
February 2010	Brenco	ETH Bioenergia	65	N/A	N/A	N/A	N/A
February 2010	Equipav	Shree Renuka Sugars	50.8	N/A	600	10.5	N/A
February 2010	Usina Vertente	Guarani (Tereos)	50	N/A	105	1.7	N/A
December 2009	Grupo Moema	Bunge	100	N/A	1,400	13.5	102
October 2009	Santelisa Vale	LDC	60	N/A	5,100	20	N/A
March 2009	Nova América	Cosan	100	N/A	N/A	10.6	N/A
December 2008	Cabrera Central Energética	ADM	50	N/A	N/A	N/A	N/A
September 2008	Agroindustrial Santa Juliana	Itochu	20	N/A	N/A	N/A	N/A
September 2008	Monteverde Agroenergética	Bunge	60	N/A	N/A	1.4	N/A
July 2008	3 greenfields in Tocantins	Bunge	100	N/A	N/A	N/A	N/A
July 2008	Usina Pau D'alho	NovAmérica	100	63	N/A	1.4	46.5
April 2008	Tropical Bioenergia	British Petroleum	50	120	60	N/A	N/A
March 2008	Eldorado	ETH Bioenergia	100	N/A	N/A	2.2	N/A
February 2008	Usina Benalcool AS	Cosanio	100	84	61	1.2	70.9
February 2008	DISA	Infinity Bio-energy	97	130	75	1.2	108
November 2007	Paralcool	Grupo Nova América	100	N/A	68	0.9	N/A
November 2007	SA Leao Irmaos Acucar e Alc.	Brazil Ethanol Participações	100	N/A	190	1.2	N/A
November 2007	Usina Conquista do Pontal	ETH Bioenergia	80	N/A	N/A	N/A	N/A
October 2007	ETH Bioenergia SA	Sojitz	33	N/A	80	1.1	N/A
October 2007	Agroindustrial Santa Juliana	Bunge	100	N/A	N/A	0.5	N/A
September 2007	Grupo Dedini	Abengoa Bioenergy Co	100	684	397	5.7	121.1
September 2007	Central Energética Paraíso SA	Infinity Bio-energy	100	N/A	52	1	N/A
September 2007	Rio Claro Agroindustrial SA	ETH Bioenergia	80	N/A	N/A	N/A	N/A
August 2007	Ibiralcool Destilaria	Infinity Bio-energy	100	N/A	28	N/A	N/A
August 2007	Destilaria Alcidia	Odebrecht	84	137	54	1.1	120.8
July 2007	Unialco	Luis & Walter Zacaner	25	N/A	10	2.1	N/A
July 2007	Andrade Açúcar e Álcool	Açúcar Guarani	68	427	206	3	144.5
June 2007	Usina Santa Luiza SA	Etanol Participações SA	100	N/A	108	1.6	N/A
April 2007	USACIGA	Clean Energy Brazil	49	213	127	1.5	138.9
March 2007	Cia Açucareira Vale do Rosário	B5 SA	51	N/A	750	N/A	N/A
February 2007	Grupo Tavares de Melo	Louis Dreyfus Commodities	100	N/A	483	5.3	N/A
February 2007	Usina Petribu Paulista Ltda	Noble Group Ltd	100	N/A	70	1.1	N/A

Source: based on newspapers.

About the authors

Marcos Fava Neves

Marcos Fava Neves is Professor of Planning and Strategy at the University of Sao Paulo, Brazil and an international expert on global agribusiness issues. He graduated as an Agronomic Engineer from ESALQ/USP in 1991, as Master of Science in 1995 and received his PhD in Management (with a focus on 'demand-driven planning and management') from the FEA/USP School of Economics and Business in 1999. He completed post-graduate studies in European Agribusiness & Marketing in France (1995) and Marketing Channels and Networks in the Netherlands (1998/1999). He specialises in strategic planning processes for companies and food production chains and is a board member of PENSA and other public and private organisations in Brazil. In 2004, he created the Markestrat think tank group, doing international projects, studies and research in strategic planning and management for more than 40 organisations. In 2008 he became CEO of Santelisa Vale Holding, Brazil's 2nd largest biofuel company. He gave more than 350 presentations in Brazil and 120 in 15 other countries. He has published 70 articles in international journals and has been editor of 25 books. He is a regular contributor for China Daily Newspaper, Folha de São Paulo in Brazil and wrote 2 case studies for Harvard Business School in 2009 and 2010.
Email: mfaneves@usp.br

Mairun Junqueira Alves Pinto

Mairun Junqueira Alves Pinto has a BA in International Relations from the São Paulo State University (2008) and is an MSc candidate in Business Administration at FEARP/USP in the area of markets and strategy. His research topics include corporate and internationalisation strategies, ethanol value chain tendencies and dynamics, and sustainability. At Markestrat he has worked on projects in strategic planning, bioenergy and sustainable business models.
E-mail: mairun@markestrat.org

About the authors

Marco Antonio Conejero

Marco Antonio Conejero is an economist and PhD candidate in Business Administration at FEA/USP. He has an MSc in Business Administration from FEA-RP/USP. In 2006 he was a visiting scholar at Howard University (Washington DC, USA) working on carbon credits and biofuels, and in 2009 at Universidad de Buenos Aires (Argentina) where he researched sustainability in value chains. As a researcher at Markestrat, he has worked on projects and publications about the analysis of business environment, strategic planning of value chains, bioenergy, green marketing and sustainability.
E-mail: marcoa@markestrat.org

Vinicius Gustavo Trombin

Vinicius Gustavo Trombin is a PhD candidate in Business Administration at FEA/USP. He has an MSc in Business from FEARP/USP. From 2005-2008 he was a researcher at pensa, where he participated in different projects about the competitiveness of agri-food chains, strategic management and corporate governance, marketing management to private and public companies, associations and cooperatives. As a consultant at Markestrat, he is involved in projects and publications about agribusiness systems in the citrus and bioenergy areas.
E-mail: trombin@markestrat.org

Keyword index